A PRACTICAL HANDBOOK ON OSTRICH HUSBANDRY

THE AUTHORS

Dr. P. Tensingh Gnanaraj is the Registrar of the Tamil Nadu Veterinary and Animal Sciences University. Being born in the southernmost district Thoothukudi (Tamil Nadu), he pursued his under graduation and post-graduation from Veterinary College and Research Institute, Namakkal. He did his Ph.D., research work in Madras Veterinary College. He has more than 24 years of research experience in the field of Livestock Production and Management. Dr. Tensingh Gnanaraj also holds a Diploma in Livestock: Technology and Farm Management from Galilee International Management Institute, Israel. For more than six years he acted as Principal Investigator of Ostrich Research Unit, Livestock Research Station, Kattupakkam. During which he established Ostrich Breeding Unit first of its kind in the country. He also improvised an inverted cycle handle bar to control aggressive birds. He also designed a hood from black cloth to restrain the Ostrich at the time of handling. He standardized package of practices for rearing Ostrich Chicks. A low-cost Ostrich egg Candler was improvised using low cost materials for effective Ostrich egg incubation. He has handled more than 36 research projects and guided more than 10 MVSc, M.Phil and Ph.D students in total. He organized more than 25 training programmes for the farmers and has more than 75 publications in International and National Journals. He was instrumental in establishing University innovation and Instrumentation center in which many farmer friendly instruments are fabricated. He has visited many countries like Israel, Australia, Brazil, China, Hong Kong, Thailand and Sri Lanka to share and gain knowledge in the Field of Livestock Production and Management.

Dr. S. T. Selvan was born in the year 1966 in the cosmopolitan city of Bangalore in the state of Karnataka, a southern state of India. He graduated in the year 1988 and did his Masters (1991) and later his Ph.D in Poultry Science (1999) from the Veterinary College and Research Institute, Namakkal, TANUVAS. He joined as faculty of Poultry Science in the year 1993 and has been elevated to the current position of Professor in the year 2009. He is currently the Professor and Head, Department of Poultry Science at Madras Veterinary College. He has more than 27 years of service in the Tamil Nadu Veterinary and Animal Sciences University, a premier University in this country. He has spent more than 20 years in various capacities at the various Research Farms of the University. He has been associated with the Ostrich Research Unit at the Post Graduate Research Institute in Animal Sciences, Kattupakkam from 2003 to 2019. During his tenure, SOP for the ostriches in the field of nutrition, breeding, reproduction, chick, grower and adult management were standardized. This intense and passionate research resulted in highest egg production, fertility, hatchability and chick livability during the subsequent years. This paved way in guiding and providing research facilities in ostriches for 10 postgraduates and 1 doctorate student. More than 28 research papers and 21 research abstracts on ostriches are published and presented in various national and International forums. The thirst for Ostrich welfare and research resulted in visiting Australia (Melbourne and Brisbane) and China (Qingdao and Hangzhou) twice and once to South Korea, Hong Kong and Italy. He has visited one of the biggest Ostrich facilities in the outskirts of Melbourne, Australia during 2000.

A Practical Handbook on OSTRICH HUSBANDRY

Dr. P. Tensingh Gnanaraj
Registrar
Tamil Nadu Veterinary and Animal Sciences University,
Madhavaram Milk Colony
Madhavaram, Chennai – 600051, India

and

Dr. S.T. Selvan
Professor and Head
Department of Poultry Science
Madras Veterinary College,
Tamil Nadu Veterinary and Animal Sciences University,
Vepery, Chennai – 600007, India

2021
Daya Publishing House®
A Division of
Astral International Pvt. Ltd.
New Delhi – 110002

ISBN: 978-93-5222-064-9

Publisher's Note:

Every possible effort has been made to ensure that the information contained in this book is accurate at the time of going to press, and the publisher and author cannot accept responsibility for any errors or omissions, however caused. No responsibility for loss or damage occasioned to any person acting, or refraining from action, as a result of the material in this publication can be accepted by the editor, the publisher or the author. The Publisher is not associated with any product or vendor mentioned in the book. The contents of this work are intended to further general scientific research, understanding and discussion only. Readers should consult with a specialist where appropriate.

Every effort has been made to trace the owners of copyright material used in this book, if any. The author and the publisher will be grateful for any omission brought to their notice for acknowledgement in the future editions of the book.

Published by : **Daya Publishing House®**
A Division of
Astral International Pvt. Ltd.
– ISO 9001:2015 Certified Company –
4736/23, Ansari Road, Darya Ganj
New Delhi-110 002
Ph. 011-43549197, 23278134
E-mail: info@astralint.com
Website: www.astralint.com

Printed at : **Replika Press Pvt Ltd**

TAMIL NADU VETERINARY AND ANIMAL SCIENCES UNIVERSITY
Madhavaram Milk Colony, Chennai - 600 051, India

Dr. C.BALACHANDRAN, M.V.Sc., Ph.D., PGDAJ., PGDEVP., DICVP, FAO Fellow, FIAVP, FNAVS, FASAW.
VICE-CHANCELLOR

FOREWORD

India with a population size of 1.3 billion and an increasing GDP of about 7% is one of the biggest consumer markets in the world. The livestock sector contributes approximately 5% of total GDP and about 27% in agricultural sector. The contribution of poultry in the livestock sector is huge in terms of population and revenue. The growth in the poultry sector is the highest in the livestock sector accounting for over 16% as per the livestock census, 2019. The growth of backyard poultry has achieved a growth of above 45%. Poultry contributes 15% of total energy requirement and 5% of total protein requirement. The total poultry market value is more than 350 billion INR (USD 8.00 billion). With a share of 6.3% and 3.3% in global egg and meat production, India ranks 3rd and 5th in egg and meat production respectively.

The gap between the per capita requirement and availability of egg (180 vs 79) and meat (11 kg vs 2.96 kg) is huge; hence, the potential for production and marketing of these products is very vivid. As per OECD, India has one of the lowest per capita daily supplies of calories, proteins and fat. It is projected that per capita chicken consumption is likely to grow from 3.2 to 9.1 kg by 2030. Similarly, the consumption of poultry along with other edible products is also poised for a growth from 12% to 24% by 2030. Chicken production (Broiler and layer) alone may not be sufficient to meet the ever-growing demands. India being in the tropical region of the world, the prevailing macro and micro climatic conditions are highly affable for poultry production. Other diversified poultry production like ducks, turkeys, Japanese quail, guinea fowls, ratites, etc should be scaled up to meet the future demands.

Ostrich production has been successfully done at Post Graduate Research Institute in Animal Science (PGRIAS), Kattupakkam since 2000. Production, reproduction and value addition of ostrich products have been carried out successfully. Commercial production in India is in nascent stage. The writing of this book is very timely and need of the hour. This book contains all the relevant and updated information on Ostrich production.

I am sure that this book would serve as a ready-reckoner for the budding professionals, field veterinarians, faculties, students and entrepreneurs.

I appreciate and congratulate the authors for their efforts in writing this book.

Date : 13.10.2020
Place : Chennai

(C. BALACHANDRAN)
VICE-CHANCELLOR

Telephone (O) : 0091-44-25551574 / 75 Fax : 0091-44-25551576

Preface

This book has been written for the sole purpose of reaching those who are interested to establish an Ostrich farm. Ostrich farming is as old as any other traditional farming, but with higher rate of return. Ratites (Ostrich and emu) are the only poultry species which have wide variety of products which are accepted universally. This book provides all the recent and relevant information required for the establishment of Ostrich farming. There have been very few authors who have written on Ostrich farming. Both the authors had been closely associated with Ostrich rearing for many years. The wide gaps felt during their days of rearing ostriches have been filled up in this book for the new comers.

Ostrich farming has been one of the money spinners in South Africa for many decades. The market for Ostrich products has not been as steady as expected in other enterprises. Initially they were reared for their beautiful flumes, then for their priceless skin and now for their exceptional health beneficial meat and oil. Lack of technical knowledge on nutrition, breeding, reproduction, health cover, etc have further led to the decline in Ostrich rearing. The content of this book has been conscientiously experienced and written. This book deals with the topics from the history of Ostrich farming to the future of Ostrich farming covering all aspects of Ostrich farming from chick to adult management, their nutrition, their health, their housing, their breeding, saleable products, etc.

The authors are thankful to the almighty, their parents and family members for having given this unique opportunity to write this book "A practical handbook on Ostrich husbandry". The authors are thankful to their teachers, thankful to the persons, organisations, etc whose names have been mentioned in this book. Information provided in this book is the understanding, personal experience and knowledge of the authors. Reference books, websites, articles, citation, etc may be consulted for further information. Finally, our thanks are due to all the persons who have directly or indirectly contributed to the contents of this book.

We are highly thankful to Dr. C. Balachandran, Hon'ble Vice – Chancellor, Tamil Nadu Veterinary and Animal Sciences University, Chennai for his guidance, encouragement and moral support for the preparation of this book

Dr. P. Tensingh Gnanaraj
and
Dr. S.T. Selvan

Contents

Chapter

1

INTRODUCTION

Ostrich has been in existence since many years. They have always been intriguing and fascinating birds for both as a farm and as a pet bird. For some it is still unbelievable, that such a bird does exist. For some, the size of egg is implausible. But this quite bird has been providing us with food, clothing, egg, utensils, feathers, etc for many years.

Courtesy:https://upload.wikimedia.org/wikipedia/commons/b/ba/Syrian_Ostrich_and_Leopard.jpg

Hunting of Ostrich has always been a sport for brave and valiant. Ostriches have featured in prehistoric rocks, scriptures, etc. Most of them have been found in and around Egypt. Records have shown that these birds have been domesticated by early Egyptians, Greeks and Romans. The Noble and Royal ladies would adorn their feathers and rode in rides driven by Ostriches. It was more for a pride than for any other purpose. In Tutankhamen's tomb, a portrayal of their king hunting the Ostriches from the chariots is more flagrantly visible. Ostriches were also used for pulling small carts. It is illustrated that the Ptolemaeus Philadelphus used a majestic chariot which were pulled by eight Ostriches. In the modern days these are totally inconceivable. It is believed that the eccentric Roman Emperor Heliogabalus had served in a dinner about 600 Ostrich brains during an important banquet.

Courtesy:https://i.pinimg.com/originals/26/2c/03/262c03a85194f9f0ffa3c26d9b84046b.jpg

Ostriches are known to have existed for more than 50 million years. Earlier the birds were much larger in size and appearances. Fossils that have been discovered suggest that they have spread worldwide from China, Southern Russia, India, Eastern and Southern Europe, Middle East and the African continent as well. The skeleton of Elephant birds has been well documented and they resemble Ostriches in real life. They had smaller head, reduced wings and long and powerful legs. The largest species had a height of over 3 m. Ostriches have virtually disappeared from other parts of the world except in South Africa. The exact reason could not be found, but they could have disappeared due to over exploitations. Currently they are found in drier parts of Africa, from the Southern Sahara to Cape Province, extending to south of Morocco, northern Sudan and southern Egypt. Ostriches have been providing human being with food, clothing, vessels and fancy feathers. Images of Ostriches have been discovered as paintings and carvings in the whole of African continent. Bushmen hunted Ostriches for their meat and adorned feathered skin for fancy purpose. Empty whole eggs have been used as drinking bowl and they are still used as soup bowl in China. Bushmen and Europeans found that Ostrich meat remained fresh for longer time and hence the meat was preferred during long journeys of either siling or hunting. During early Sparta days, eggs were regularly used as an exhibition item. Egg shell fragments were used as an ornamental wear for the folks of tribal. In West Africa, it was considered to possess magic properties, protecting Muslim houses against lightening. Ostrich eggs on graves in South Ethiopia denote number of enemies killed by the deceased. Hunting of Ostriches was considered as a privilege game of the Pharaohs. In Tutankhamen's tomb, there seems to be a depiction of king hunting an Ostrich with a bow and arrow from his chariot. Tutankhamen had a long handled golden fan which held Ostrich feathers and is decorated with an image of a Pharaoh hunting an Ostrich. The capture of Ostriches is depicted in the mosaic of the corridor of the great hunting scene in the famous fourth century villa of Piazza Armenia in Sicily.

Courtesy: https://i.pinimg.com/originals/ec/1c/1b/ec1c1b661224107b2d1cb00b666696dc.jpg

Assyrian Kings are seen strangling and slaughtering Ostriches on regal seals dated from the 8th century and they also considered them as holy. Ostrich egg and feathers are featured in Egyptian hieroglyphs. Display in British museum shows an Ostrich being presented to a Pharaoh. Ostriches have been referred to on many occasions in Bible. Solomon saw the symmetrical Ostrich feather as a symbol of justice. Egg shells are used as ornaments in Coptic and Greek Orthodox churches. Greek and Roman Generals decorated their helmets with Ostrich feathers. Ostriches were encountered by the Chinese during military missions and their images are featured in sculptures on the sides of tombs of emperor. Crusading knights returning from the Middle East introduced the Ostrich feather into Western Europe where it became fashionable with the monarchy. The coat arms of the British Prince of

Wales have three Ostrich feathers, an emblem established in the 14th century. The skin of Ostrich has been used in protective jackets in the Arab world. Other than the aboriginal hunting, Ostrich meat has been largely ignored.

Ostrich eggs have been found to be traded some 5000 to 6000 years ago. They have been a subject of debate, whether the Ostriches are wild or domesticated. Since, the trading is usually done for the precious items; it is understandable that eggs were very precious at that time. Empty and whole eggshells have long been used as storage and drinking vessels in Africa and Arabia. Eggshell fragments are present in as a fashion in jewellery used by bushmen and surprisingly this continues even to the present day. Ostrich has also found its name in Bible. It is termed as unclean meat in Old Testament. The Ancient Egyptian, Solomon saw the symmetrical Ostrich feather as a sign of justice. Eggshells were also used as ornaments in Coptic and Greek Orthodox churches, and in West Africa they are used to protect Muslim houses from lightning. The Jews refer to Ostriches as being possessed with Supreme powers, so that they can hatch their eggs with a single "look" after seeing the sky.

The Ostrich is sometimes referred to as 'Camel bird' because of its similarities with dromedaries and it was named in 1758 by Linnaeus as Struthio camelus, based on the Greek and Latin name Strutho camelus. The birds have a walk similar to the camels, and as dessert birds, they have the protecting third eyelid which is not present in chicken.

Courtesy: https://mizanproject.org/wp-content/uploads/2019/05/BirdonEggsCrop.jpg

World Ostrich production

South Africa is the leading country in Ostrich production. Europe has a very good market for Ostrich meat due to its high protein and low-fat content and it is available through the supermarket chains. Eastern and South Eastern countries are a good market for Ostrich feathers and skin. The table below reveals the number of Ostriches slaughtered during 2005 and 2006.

Table.1 Ostrich slaughter (2005 and 2006)

Country	2005	2006
South Africa	1,50,000	2,00,000
China	50,000	100,000
Zimbabwe	20,000	20,000
Australia	12,000	22,500
Brazil	10,000	50,000
USA	10,000	13,500
Israel	10,000	10,000
Hungary	8,000	12,000
Spain	7,000	7,000
Philippines	5,000	7,500

Source: World Poultry, 2006

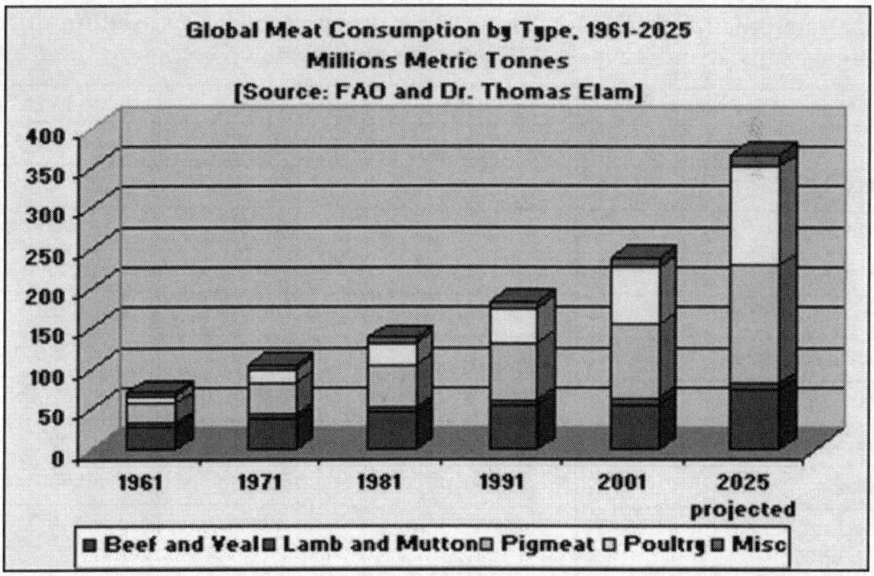

Courtesy: https://world-Ostrich.org/wp-content/uploads/2012/11/30fig1.gif

The above projections suggest that poultry meat is likely to be in great demand during 2025. To fulfil this demand, Ostrich can play a major role in achieving, as the meat production can be speeded up if it demands.

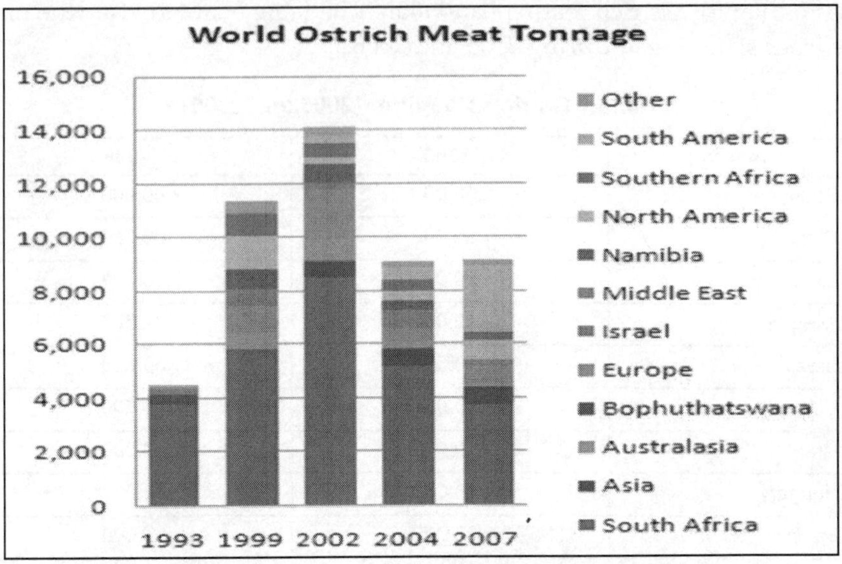

Courtesy: https://world-Ostrich.org/wp-content/uploads/2013/03/ostmeat-ton08-1.gif

Even though South Africa is the leader in Ostrich production, other nations like China, Iran, Brazil, Pakistan, etc are catching up with their increased production. A more recent statistics reveals the fact. The same is shown below;

Name of the country	2000	2010	2018 - 2019
Brazil	0	450000*	250000*
China	250000*	500000*	500000*
South Africa	300000	250000	130000**
Pakistan	0	0	100000*
Iran	0	0	40000*
Arabian countries / Emirates	0	0	250000*
Botswana	0	0	15000
New Zealand	10000	15000	15000
Australia	30000	15000***	15000***
Israel	25000	1000	0
Namibia	25000	2000	0
Zimbabwe	55000	5000	0
Ukraine	0	1500*	50000****
Romania	0	1000	10000
Poland	0	5000	3000
Germany	1000	1750	2500

Portugal	2000	2000	2000
Hungary	0	1000	1500
France	500	1500	1500
Austria	500	1000	1000
Bulgaria	0	0	1000
Italy	5000	2000	1000
Spain	7000	1500	1000

*For home respectively, regional market only **temporarily banned for export, ***Export into the USA and Japan only, ****Not yet approved for EU

Courtesy: http://www.ltz.de/en/news/lohmann-information/2019-1/3_Ostrich_Production_in_Germany-ecological_way_to_economic_success.php

❑❑❑

Chapter 2

HISTORY OF OSTRICH FARMING

Whenever we pursue to find the antiquity of Ostrich farming, the first and only country that emerges to tell us about the history of Ostrich farming is South Africa. It has been recorded that history of Ostrich farming is same as history of Ostrich farming in South Africa. So, in a nutshell, the information about the past, present and future is held with South Africa. South Africa could play a very vivacious role in the future development of Ostrich as an industry. Currently, the country holds more than 80% of total Ostrich population of the world.

Jan van Riebeeck documented large flocks of Ostriches roaming in Cape during 1652. During 1821, Ostriches were hunted for their plumes (feathers). In 1822, Cape colony made a proclamation to protect wild Ostriches from being hunted to extinction. The first wild Ostrich was tamed in 1863. A severe drought during 1865, prompted the farmers to rear Ostrich as a supplementary income. A Cape Colony census reported 80 domesticated Ostriches in that year. Lucerne was introduced during 1869by the then Magistrate of Oudtshoorn, Mr. Scholtz who imported the seeds and planted in a small plot to feed the Ostriches. The same year the first prototype of Ostrich incubator called "The Eclipse" was also introduced. The feather auction was first started during 1870 in Mossel Bay. Ostriches were imported from the North Africa in 1874 to improve the feather quality through cross breeding. According to Cape Colony census, there were 32,247 farmed Ostriches during the **first boom (1875 – 1880)**. In 1882, the export value of Ostrich feather reached its peak. In 1884, the local government imposed heavy duty to protect the local industry. A year later in 1885, the industry collapsed under its own weight of heavy production and poor feather quality. In 1886, a decision was made to import Ostriches from Barbary (North African red neck). Similar decisions were made during 1888 to 1903 to improve the local strain (South African Blue neck). By 1890 strict breeding programs including selection, cross breeding and rigid culling to improve the local strains was introduced. During 1885, there were 2,53,463 domesticated Ostriches as per the census record of Cape Colony. **The second boom started during 1897**. Most of the Ostrich farms were concentrated in Oudtshoom, Grahamstown, Cradock and Graaff Reinet with Port Elizabeth the headquarters of feather business. Land prices increased significantly and feather "palaces" were built. The farmers were very wealthy during these times. The Lucerne industry also developed along with feather industry. The second Anglo – Boer war (1899 – 1902) had an indirect effect on this industry. Feathers could not be moved to auction area during war times. The feather industry improved after the end of war. In 1907, a law was passed that made illegal export of Ostrich punishable by a jail term.

Courtesy:https://upload.wikimedia.org/wikipedia/commons/d/d8/ Dyeing_of_Ostrich_feathers_1919.jpg

In 1913, Ostrich feather was the fourth largest export item from the Union of South Africa after gold, diamond and wool. During the same year, the government passed a law to prevent extinction of exotic and rare birds. This affected the practice of Ostrich industry. The outbreak of World War I had tremendous effect on this industry. This was the end of feather industry. The war time affected export, logistics and related activities. Feather fashion fizzled out. An estimated 80% farmers were out of business and livelihood.

The "*Suid-Afrikaanse Volstruisboere Kooperasie Beperk*" was established in South Africa to develop the industry, regulate the market and stabilise the prices. The organisation failed miserably and the Ostrich population dwindled from one million to 23,000 in 1930. In 1940's the interest for the Ostrich leather rejuvenated the growers and led to resurgence of the Ostrich enterprises. The Klien Karoo Agricultural Co-operative was established in 1945 to regulate the industry and also to set the quality standards. The industry was protected as a national asset. The first abattoir was opened in 1950 and the same was taken over by the government in 1958. Ostrich meat became popular and the dried meat called "biltong" was very successful in marketing as red meat. This meat became a credible competitor for red meat as a whole. The KKLK society opened their own abattoir in 1964 and subsequently the tannery in 1970.

Courtesy:https://investoramnesia.com/wp-content/uploads/2019/02/Ostrich-bubble1024x621.jpg

The Ostrich industry was de-regulated in the year 1993 and it resulted in free market system. During 1993, first abattoir was built specifically to export meat to Europe. The industry had to meet the challenges from other regions including Australia, US and few European countries. The establishment of South African Ostrich Business Chamber (SAOBC) supports the Ostrich industry till now. It played very significant role during the difficult times of Ostrich business. It represents all the role players in the Ostrich industry. The Western Cape Department of Agriculture provided all the research support since the establishment of the Oudtshoom Research Farm during 1964. During the outbreak of Avian Influenza in 2004 and 2011, the industry faced major challenges and could mitigate their effects with implementation of strict protocols. This proved that the Ostrich Industry in South Africa is no stranger to such boom and recession and has always proved to be resilient. The industry further suffered due to volatile exchange rates, since the industry was dependent to the extent of 90% on foreign exchange. The South African growers still market their leather products through innovative marketing strategies. And simultaneously they have developed local market for Ostrich meat.

Landmark events in Ostrich farming in South Africa

1838 – Export of Ostrich feathers to the European countries

1857 – First successful hatching in Algeria

1863 – Establishment of first organized farm in Klein Karoo Landboukooperasie and Eastern Cape region of South Africa for marketing of Ostrich feather

1869 – Invention of Ostrich egg incubator by Arthur Douglass

1870 – An Act of preservation of the wild Ostriches was passed in Cape Colony, South Africa

1884 – Large shipments of Ostriches were moved from Africa to America and Australia

1885 – Major slump in Ostrich marketing

1925 – Establishment of first cooperative society

1945 – Establishment of first modern and second cooperative Ostrich farm, the Klein Karoo Landboukooperasie (KKLK) at Little Karoo to regulate feather sales

1964 – Establishment of first Ostrich abattoir for mass production of Ostrich meat

1969 – Establishment of first leather tannery in Oudtshoom

1994 to 1998 – Deregulation resulted in spreading Ostrich activities from Klein Karoo region to other parts of Africa. This period also witnessed formation of many National Associations like the Ostrich Breeder's Society of South Africa, the South Africa Ostrich Producers Association and the South African Ostrich Processors Association

1998 – A small number of young Zimbabwean blues (S.c.australis) were introduced into the

1999 – 2010 Various research programs were undertaken on hatchery, reproduction, leather quality and nutrition of Ostrich.

2012 – Researches on breeding, genetics selection, cross breeding, nutrition, etc were undertaken. Development of various practical models for Ostrich nutrition, incubation and hatching of eggs were attempted.

Origin and development of Ostrich farming as chronicled in Oudtshoorn Research Farm: Celebrating 50 years of the world's first Ostrich Research Farm (1964 – 2014)

At around 1863, Ostrich farming became a viable agricultural undertaking after the introduction of wire fencing and lucerne farming in South Africa. The then magistrate of Oudtshoorn, Scholtz, introduced lucerne when he imported seeds which he planted on a small plot to feed his Ostriches. By 1870, Ostrich farming had become an extremely profitable industry, aided by the invention in 1869 of an Ostrich egg incubator by Arthur Douglass, the self-described pioneer Ostrich farmer of Grahamstown. Douglass also identified the Ostrich wireworm and wrote the first book on Ostrich farming, published in 1881 entitled "Ostrich Farming in South Africa".

Many a time at first, we were told we were mad, and should leave it alone; that it would never pay; that the birds were naturally so timid a nature, they would never breed in confinement... Extract from Ostrich Farming in South Africa, 1881.

More informationat :

http://www.elsenburg.com/sites/default/files/attachments/201611/OUDTSHOORN%20RESEARCH%20FARM%2050%20YEARSLOW%20RES_FINAL.pdf

Development of Ostrich farming in Australia

Ostriches were first imported into Australia in 1868 or 1869 when the Zoological Acclimatisation Society of Victoria acquired five immature birds from the Cape Colony. Of these, were males and three were females, one of which proved to be barren (Anon 1873). The birds were moved to the Wimmera under the care of Mr Samuel Wilson who harvested feathers which were sold at high prices. The South Australian pioneer pastoralist, T.R. Bowman first attempted to farm Ostriches in the mid 1870's (Iwanicki 1985), and by 1904 was running nearly 500 birds at Campell House Station. In 1881 W. Malcom established the Barossa Ostrich farm with 7 adult birds from M. Bowman of Port Wakefield and 3 adult and 13 young birds from South Africa (Herbert 1913; Iwanicki 1985). Later, W. Malcolm formed `The Malcolm Ostrich Company' in partnership with Francis Bignell at a site north of Port Augusta. The company flock was enlarged by the importation of a further 94 birds from South Africa, but soon after, in 1885, the partners sold their business (Iwanicki 1985). It then became the 'South Australian Ostrich Company' with the major shareholders being J.B. Cudmore, J.H. Angus, T.R. Bowman and J.M. Anderson. The company was formally wound up in 1916 and almost three hundred birds were moved to Priors Farm at McLaren Vale south of Adelaide. In 1926 all the stock were sold to a circus (Iwanicki 1985), marking the end of an important facet of Ostrich farming in Australia. Fortunately however, birds were purchased contemporaneously by other potential Ostrich farmers throughout Australia. By 1913, J. Barracluff of South Head, Sydney had one hundred birds. In 1900, a pair of Ostriches was placed at the Hawkesbury Agricultural College and their feathers won a gold medal at a Franco-British Exhibition. In 1905, Cairnes of Nardoo, and Sanderson of Gilgandra started a company with 6 birds purchased from the South Australian Ostrich Company. In 1907 the flock was moved to Coonamble and by 1913 had increased to 550 birds. A very important flock in Australian Ostrich history is the one at Yanco. In 1913 this flock contained 90 birds. Among these were 12 Sudanese Ostriches which had been imported from North Africa in 1912. At the Yanco Experiment Farm these were mated with birds of the South African strain purchased from South Australia (Herbert 1913) to improve feather quality. After the 1914 slump the Ostrich disappeared from Australian agriculture to zoos, wildlife parks and some were kept as pets. The ruined buildings and fences of the Port Augusta farm are still visible on the western plains of the Flinders ranges. The current owner is farming the remaining feral Ostriches which are descendants of the original flock (Iwanicki 1985).

In the late 1970's and early 80's a number of farmers became interested in the potential of farming Ostriches. Procedures for farming and breeding were not well known and information was hard to find, thus flock numbers increased very slowly. In 1988, the Australian Ostrich Breeders Association was formed and now has members successfully breeding birds in New South Wales, Victoria, South Australia, Western Australia, Queensland and Tasmania. In 1990 it had a membership of 113. Eighty-five per cent of members are in Victoria and New South Wales. R.Elliott (pers. comm. 1991) estimates that there are presently approximately 1870 birds in Australia. There are many factors responsible for the slow rate of growth of the national flock, among them, reproductive failure.

There are presently only small numbers of birds within Australia and these are descendants of those imported around the turn of the century. Due to current quarantine regulations and facilities, birds cannot be imported; hence growth of the industry is dependent upon existing stock. This will extend the time necessary to breed sufficient numbers of birds to form a viable commercial industry. R. Elliott (pers. comm. 1991) estimates this stage may be reached by 1997. The effects of the limited genetic base on population growth are not known at present but it could have implications for reproductive efficiency. At the present time the birds have a very wide geographical distribution. This is not currently a problem but may become one if the industry aims to integrate its functions. If, for example, it proves desirable to develop a fully specialised processing facility for Ostrich products, it would then be necessary for the industry to be centred in a smaller area to reduce transport costs and to increase efficiency. The exact location rests on environmental issues and cannot be decided until studies on climatological requirements for successful breeding of Ostriches have been made. At this stage production levels in different climatic areas are similar, but bird numbers are small and practices vary greatly between farms, making comparisons extremely difficult. It will be some time before the Australian Ostrich Industry becomes fully commercial. The problems identified in production such as egg infertility, embryonic mortality and post hatching leg deformities must be dealt with while bird prices are high and there are resources to fund research. Much needs to be done to develop a viable secondary industry. Issues such as quality control, processing requirements and facilities, centralisation, market research and an industry plan for the future all require definition and development. The Australian Ostrich industry has a great deal of potential to become a substantial export industry for Australia. Its success is going to depend upon solving some of its present problems and developing substantial guidelines for future growth. It is important to have a central organisation to co-ordinate these developments.

More detailed information at:

http://livestocklibrary.com.au/handle/1234/19657

Development of Ostrich farming in USA

Ostrich farming is in its infancy in the United States, the first Ostriches having been imported in 1882. Other importations followed, but it was some time before success was obtained in the reproduction of Ostriches in this country. The first Ostrich breeders in the United States learned through experience alone, as there was available very little information on Ostrich breeding and management and no data concerning their care and management as adapted to climatic and soil conditions in this country. From these importations the Ostrich farms have been built up and breeders have located in various parts of the United States, so that to-day there are Ostrich farms in Arizona, Arkansas, California, Florida, and Texas. Complete statistics of the number of Ostriches in this country are not available, but reports received from all of the large and the majority of the smaller Ostrich farms show that there were at least 6,100 breeding or feather-producing Ostriches in the United States in January, 1910. These Ostriches were distributed among the States approximately as follows: Arizona, 80 per cent; California, 17 per cent; Arkansas, 2 per cent; while Texas and Florida together reported less than 1 per cent of the total number in this country.

More information at:

https://www.google.co.in/url?sa=t&rct=j&q=&esrc=s&source=web&cd=&cad=rja&uact=8&ved=2ahUKEwi8xdfE7NnqAhXzTgGHZtDCuoQFjAAegQIAxAC&url=https%3A%2F%2Fnaldc.nal.usda.gov%2Fdownload%2F5420953%2FPDF&usg=AOvVaw3CK6MxhyK7zi-lGJjl5o0Y

According to the American Ostrich Association, there is no ingle reliable comprehensive information on production, price, and market of Ostrich. The 2012 census of Agriculture notes that 258 farms raised 6540 Ostriches and marketed about 3141 numbers. This is a dramatic decrease from 2007, when 714 farms raised 11,188 Ostriches and marketed about 5697. This continues decline from 2002. Texas, followed by California and Kansas are the top three Ostrich producing states.

More information at;

https://www.agmrc.org/commodities-products/livestock/Ostrich-and-emu-53585

Development of Ostrich farming in Poland

The first Ostrich farm was located in the north of the country in Garczyn near Gdndsk during 1993.In 2007, there were 180 farms with a total population of more or less 14,000 including 1800 breeder birds. Poland had become leaders in Ostrich farming along with Italy and Spain. Since 1993, the Ostrich development can be distinguished in three phases. The first phase involved reproduction (1993 -2000), second phase was mixed phase including reproduction and commercial farming (2001 – 2004) and third phase was commercial production from 2004 to 2008.

More information at:

https://www.researchgate.net/publication/233647137_Ostrich_farming_in_Poland_Its_history_and_current_situation_after_accession_to_the_European_Union

Development of Ostrich farming in China

Ostrich (*Struthio camelus*) has become one of the valuable agricultural enterprises in many countries of the world. China has become Asia's number one Ostrich species raiser with about 20,000 Ostrich bred farms. China started in 1992 to import some Ostrich species from South Africa. China's progress in producing Ostriches has improved its rank from fifth to the second globally in recent years, due to the following factors: The area of pastures which reached 40% of the area of China's large amounts of agricultural wastes. China has been producing iron and thus has produced the cheapest agricultural machines, such as irrigation machines, cutting and pressing the feed. China achieved progress in the electronic industries which have enabled it to produce hatchery machines to promote the poultry industry. China is making use of solar energy and producing cheap energy used to irrigate agricultural land and operate farm machinery. Thus, the price of feed is cheap and therefore it costs to feed the Ostriches as little as possible. China has imported large amounts of Ostriches from South Africa and used the best breeding and Ostrich production systems in South Africa. The most important factors leading to the success of the Ostrich industry in china is limited into three factors. First: Nutrition and cheap feed by planting enough alfalfa using solar energy to irrigate these lands, sowing saline and drought tolerant genotypes, such as blubanks and bonecam, as well as machines for cutting and making pellet for green feed. Second: Hatching Third: Health.

More information at:

https://www.researchgate.net/journal/11100338_Zagazig_ Journal_of _Agricultural_ Research

Development of Ostrich farming in Nepal

Ostrich Nepal is the private company which started the farming of Ostrich in 2008. It is the largest single Ostrich farm in Asia, which is located at Tilotamma municipality, Rupandehi covering the land of Twenty two (22) bigha(8.81 Acers) and in Suryapur VDC, Rupandehi covering a very big mass of land of one hundred and fifty(150) bigga (60 Acers). With the increasing demand of the farmers and also to give the production a bit of commercial and scientific upgradation, this farm has also established itself in Bijauri, Dang where parents Ostrich are kept covering a total farm land of altogether 50 bigga. This farm has been through from egg production to hatching eggs to meat packaging and customer delivery. It is distributed in the market through various outlets and the Ostrich dishes are available in several restaurants of the city.

More information at:

https://www.google.com/url?sa=t&rct=j&q=&esrc=s&source=web&cd=&cad=rja&uact=8&ved=2ahUKEwi5rfvknN7qAhXk4zgGHRw2DK0QFjALegQIARAC&url=https%3A%2F%2Fwww.Ostrichnepal.com%2F&usg=AOvVaw1A9i7X5UBwr3HW4TPRVmb&cshid=1595330385441671

Development of Ostrich farming in Iran

In the recent decades, Ostrich farming has been developed in different countries across the world, including Iran, where the commercial Ostrich farming began in 1999. Nowadays, this industry is spreading all over the country, especially in the central areas (Khodaei Motlagh, 2015). The growing trend of this industry indicates its compatibility with the weather and climate in different parts of Iran. Different studies on Ostrich farming revealed that it is economically beneficial (Kohansal et al., 2014). Industrial Ostrich farming is extending globally due to its high-quality leather, low-cholesterol meat, egg, eggshell, and oil.

With the production between 50,000 and 60,000 Ostriches per year, from which 3-3.6 million kilograms of meat are obtained, Iran is ranked the second biggest producer of Ostrich meat in the world after South Africa, the head of Ostrich Producers Association said. "Meat is not the only product derived from this precious bird. Its feathers, leather, oil, egg and eggshells also bring in profits and create value added when processed," Behrouz Rezaei also told Financial Tribune. Ostrich farming on the industrial scale, he added, began in Iran around 14 years ago. But in the first few years and due to lack of sufficient expertise, most of these ventures failed and businesses stopped expanding.

Source:

https://www.google.com/url?sa=t&rct=j&q=&esrc=s&source=web&cd=&cad=rja&uact=8&ved=2ahUKEwjZvLGn97qAhUezDgGHZz3ARQQFjAOegQIAxAB&url=https%3A%2F%2Ffinancialtribune.com%2Farticles%2Feconomy-domestic-economy %2F87818%2Firan-ranks-world-s-second-biggest-Ostrich-meat-producer&usg = AOvVaw2HS68aE9JTkcEOEw8Bj2dB

Chapter

3

CLASSIFICATION AND FEATURES OF OSTRICH

Ostriches which belong to the Ratite family are known to have existed for more than 50 million years. The Ostrich (*Struthio camelus*) belongs to an ancient group of birds and the order of flightless birds called ratites. Other ratites include the Emu (Australia), Rhea (South America), Cassowary (New guinea papau Islands)) and Kiwi (New Zealand). Earlier the birds were much larger in size and appearances. Fossils that have been discovered suggest that they have spread worldwide from China, Southern Russia, India, Eastern and Southern Europe, Middle East and the African continent as well. Ostriches have virtually disappeared from other parts of the world except in South Africa. The exact reason could not be found, but they could have disappeared due to over exploitations. Currently they are found in drier parts of Africa, from the Southern Sahara to Cape Province, extending to south of Morocco, northern Sudan and southern Egypt.

The following species have been described from the fossils remains;

- *Stuthiokaratheodoris* – found in the Islands of Samoa
- *Struthio novorossicus* – found on the west coast of Black Sea
- *Struthio odlawayi* – found in Algeria, Egypt and East Africa
- *Struthio asiaticus and S. indicus* – found near river Kain in India
- *Struthio wimani* – found at T'ai Chia Kau in Pao Te Hsien, Nporth west Shansi on the Yellow River, China

The above species were much larger than the current species. There is remarkable similitude between the Ostriches and the extinct Elephant birds of North Africa and Madagascar (family Aeopyornithidae, genusAepyornis).

Taxonomic Classification

Phylum	: Chordata
Sub phylum	: Vertebrae

Class : Avis

Order : Struthioniformes

Sub order : Struthiones

Family : Struthionidae

Genus : Struthio

Species : Struthio camelus

Sub species (RACES):S.c.camelus, S.c.molybdophanes, S.c.massaicus, S.c.australis, S.c.syriacus, S.c.spatzi,etc.

Description Of Common Ostrich Subspecies

Subspecies	Location	Male skin	White neck collar	Bald crown	Tail feathers	Iris colour
S.c.camelus	South Morocco, North Uganda	Pinkish	Yes	Yes	White	Brown
S.c. molybdophanes	North East Ethiopia, Somalia	Blue grey	Broad	Yes	White	Blue
S.c.massaicus	East Kenya, North Tanzania	Pinkish grey	Narrow	Absent	White	Brown
S.c.australis	North Namibia, Zimbabwe	Grey	Absent	Absent	Brown	Brown

Commercial Classification Of Ostrich Sub-Species

Commercial Name	**Sub-species**
Red Necked	Struthio.c.camelus
	Struthio.c.massaicus
Blue necked	Struthio.c.molybdophanes
	Struthio.c.australis

Is there a term like domestic Ostrich?

The domestic Ostrich (Struthio camelus var. domesticus) is technically a variety and not a subspecies by itself. Moreover, technically it is a hybrid of several wild species. The South Africa Ostrich (Struthio camelus australis) is the foundation stock of the present domestic Ostriches. The terms used to describe "African black" or "Black necked Ostrich" are coined by American Ostrich farmers as trade names to be used in local area. The domestic Ostrich maintains the appearance of the original stock. But compared to all wild species, it is generally smaller with shorter legs and shorter mandibles. The distinguishing feature is the feather quality, where the barbs are broader than the wild types and the tips are rounded rather than pointed. During the late 19th century, several breeder stocks from North Africa and Arabian Ostriches were brought to South Africa and were selectively bred using the intense selection method. They concentrated mainly for the feather quality.

General Characteristics of Ratites

Parameters	Ostrich	Common Rhea	Emu	Cassowary	Kiwi
Body size					
Height (m)	2.5	1.5	1.6	1.5	0.4
Weight (kg)	135	25	38	38	2
Male	Black	Grey	Grey-black	Glossy black	Tan-brown
Female	Grey-brown	Monomorphic	Monomorphic	Monomorphic	Monomorphic
Toes	2	3	3	3	3
Speed (km/h)	70	50	50	50	45
Egg colour	Creamy white	Golden white	Glossy green	Green	Glazed white
Egg wt (kg)	1.5	0.6	0.6	0.65	0.45
Incubation period (d)	42	42	58-61	30	75-78
Incubating sex	M/F	M	M	M	M
Natural distribution	Africa south of the Sahara	Brazil, Bolivia, Argentina	Australia (except east)	NE Australia, New Guinea	New Zealand

Any commercial venture should have a set goal and the work should be towards that goal. Under commercial farming the goals or the production parameters should be strictly adhered to, so that the expected returns can be achieved. Ostrich farming may have an exception as the commercial exploitation of this farming is yet to be totally understood. However, there are some basic production parameters that have to be adopted so that the farming is viable and sustainable.

Production (Commercial) Traits Of Ostriches

Life Span	:	70 – 80 years
Production life	:	30 – 40 years
Market age	:	12 – 14 months
Market weight	:	100 – 120 kg
Adult weight	:	120 – 180 kg
Adult height	:	1.8 to 3.0 m
Sexual maturity	:	Females (24 –36 months) Males (36 – 48 months)
Egg production/year	:	50 to 60 nos.
Breeding season	:	Dec / Jan to July / Aug
Fertility	:	50 – 60 %

Egg weight	:	1300 to 1500 g
Incubation period	:	42 - 44 days
Chick weight	:	750 to 900 g
Growth rate in chicks	:	280 to 300 g (up to 3 months)
No. of chicks per year	:	10 to 20 per female
Mortality percentage	:	40 - 50 % in young stock 1 – 2 % in adult stock
Major products and yield	:	Leather : 14 to15 sq.ft per bird Meat : 50 to 55 % (50 to 60 kg)
Minor products and yield	:	Oil : 4 to 6 litres per bird Feather : 900 to 1000 g per year per bird Egg shells : 20 to 30 nos.

Growth parameters:

1. Chick weight
2. Chick liveability
3. Growth at regular intervals (monthly)
4. Mortality in chicks, growers and adults
5. Feed consumption
6. Feed conversion ratio

Reproduction parameters

1. Age at sexual maturity
2. Age at 25% egg production
3. Age at 50% egg production
4. Hen day egg production
5. Hen housed egg production

Hatching parameters

1. Fertility percentage
2. Egg weight loss percentage
3. Hatchability percentage (on total egg set and fertile egg set)
4. Economics of meat production
5. Economics of egg production

There seems to be lots of anatomical differences among the ratite species. The various differences are listed below;

Anatomical differences among the three common ratite species

Features	Ostrich	Emu	Rhea
Body size	**Largest**	**Second largest**	**Smallest**
Proventriculus	Very large, located dorsal and directly caudal to the ventriculus, with a dorsal strip of glandular tissue.	Small with glands encircling the cranial 2/3 to 3/4 of the organ.	Small dilation, with a nodular gland on the dorsal 1/3 of the organ.
Ventriculus	Large, round and muscular with a restrictive pyloric valve.	Smaller, round and muscular with a less restrictive pyloric valve.	Elongated, expandable and muscular, with a less restrictive pyloric valve.
Large intestine	Very long, with proximal sacculated and distal non-sacculated segments and long colon	Very short; non-sacculated.	Very short; non-sacculated.
Apterylae	Lateral thorax and abdomen, ventral wing, below the vent and legs.	Ventral wing, and leg below the hock joint	Ventral wing and leg below the hock joint.
Toes	The greater toe has four phalanges the lateral has five phalanges	Three toes, medial toe has three phalanges, central toes have four, and the lateral toe has five.	Three toes, medial toe has three phalanges, central toe has four, and the lateral toes have five.
Male and Female	Dimorphic: male-black and white feathers, usually larger body size. The beak, lower legs, and skin turn bright pink to lavender during breeding season. Female - grey and white feathers.	Monomorphic - charcoal grey. Body size is not a distinguishing feature.	Monomorphic – white or grey and white, although the male has a distinctly larger body size.
Tail	Mobile and functional, with eight vertebrae caudal to the illium, the first six are always free and last two may be fused together.	Immobile, indistinct, consisting of eight vertebrae caudal to the illium, the first five are free and the last three are fused together.	Small and slightly with mobile vertebrae caudal to the pubis.
Wings	Large.	Short and vestigial.	Large.
Jugular vein	Only the right is present.	Both right and left are present.	Right is always present; left is absent in some individuals.
Pubis	Caudal ends are in close proximity and curve forward to form a pubic floor. Fused to the ischium.	Caudal ends are free and widely separated.	Caudal ends are widely separated and fused to the ischium.
Caeca	Long, with a large lumen, and sacculated.	Short, small lumen and non- sacculated. well developed	Long, with large lumen, and sacculated.

Features	Ostrich	Emu	Rhea
Clavicle	Absent.	Present.	Fused to thoracic girdle.
Tracheal Diverticulum	Absent.	Present.	Absent.
Skin on metatarsus	Grey / tan to pink in color, cranial scutes.	Black in color, cranial scutes and raised caudal scutes	Gray in color, cranial scutes.
Sternum	Bilaterally fused together.	Bilaterally fused together.	Becoming partially fused in adults.
No. of Cervical Vertebrae	18	18	15
No. of complete Ribs	5 or 6	5	3
No. of Renal Portal Valves per Kidney	3	1	1
Testes	Paired, tan in color, similar in size.	Paired, green in color, left larger than the right.	Paired, tan in colour, similar in size.
Phallus (male)	Long, large, solid with no ostium.	Short, cartilage base, with ostium, containing invaginated phallus.	Short cartilage base with long, non-erect phallus with ostium, containing invaginated phallus.
Gall Bladder	Absent.	Present.	Present.

EMU

OSTRICH

EMU

CASSOWARY

KIWI

Chapter

4

SYSTEMS OF MANAGEMENT

As usual, the dilemmas always exist as to how much area is required to start an OSTRICH farm. The requirement of area to start an Ostrich farm would largely depend upon the availability of capital, availability of inputs, capacity to produce and maintain the birds, marketing potential, etc. Unlike other poultry farms, Ostrich farming is more complicated because the standard features and requirements are not available and also the birds are still primitive and not totally domesticated. They have more of "jungle/ wild instinct" and in spite of domestication for more than a century, they have not been totally domesticated (only semi domesticated). The choice as to which system to use is essentially governed by availability of land, scale of production and labour and feed costs. However, the following systems exist for Ostrich management worldwide; In the first two systems, there is an additional choice as to whether to use natural or artificial incubation. With intensive systems, artificial incubation and hatching of eggs are always employed

1. Extensive system with natural hatching
2. Extensive system with artificial hatching
3. Semi-intensive system with natural hatching
4. Semi-intensive system with artificial hatching
5. Intensive system

Extensive system with natural hatching

Ostriches basically love open space. Under the present conditions, it is not feasible to provide large area to Ostriches. Due to the current population boom along with other industrial activities, sufficient space for Ostriches is not possible. But, wasteland unfit for any agricultural purpose can be totally utilized for Ostrich farming. But larger area would also cause monitoring of birds unpractical. Such system is most suitable for "ranch" type of farming, which is rarely seen in India. The main and only advantage of this system is the least cost of maintaining the birds and hatching of eggs. No scientific data can be obtained from such flocks. The exact number of birds available in the area also cannot be accounted. Disadvantages of this system are high unaccounted mortality in wild conditions of the young and old ones, capture of chicks for examination is not possible, unhealthier chicks in the wild conditions, etc. In total, this system is not economically viable. This system is more suitable to Zoological parks, where the birds can be purely maintained on low input cost with natural feeding and breeding.

This type of system requires a large area of land in excess of 40 ha (100 acres). Apart from the cost of the birds (which is common to all systems), land is the major capital requirement. The birds are kept and raised as near as possible to their natural habitat, with minimum interference. The main advantage of the extensive farming system is the greatly reduced cost of keeping adult birds in large numbers.

However, the disadvantages can easily outweigh the advantages of this system. Monitoring and identification of the birds and collection of eggs (for artificial incubation) often prove to be a problem. No control over breeding conditions can be exercised. Eggs collected for artificial incubation may be transported over rugged land and this may often cause damage to the delicate internal structure of the eggs. Mortality and loss of birds are often high, particularly among chicks, because of the high degree of predation. The physical capture of birds is complex and expensive, and birds are generally, difficult to handle through lack of human contact.

Extensive system with artificial hatching

Under this system of management, Ostriches are "maintained" in wild conditions; eggs are collected and hatched in the incubators. The chicks are then raised under supervision and after 3 to 4 months, they are let in the wild conditions. This system requires some basic housing for maintaining incubators and raising the chicks. This system is feasible if large wild area is available for farming. The disadvantages observed in the previous system are also observed here. In addition to the above, care of hatching eggs will be minimum, as identification of nests, collection of eggs and time of laying eggs would never be known.

Semi-intensive system with natural hatching

This system also requires at least 20 to 60 ha (50-150 acres). In a way, the birds should have a feel of openness. The birds are kept in relatively small paddocks or territories of approximately 8 to 20 ha (20-50 acres). They are able to roam freely to a certain extent, thus obtaining some of their nutritional requirements from the pasture. Supplement feeding (amounting to 40-60 percent of total feeding) is required to ensure that all the birds' nutritional requirements are met. Feeding sites should be located near the perimeter fencing, to increase accessibility and reduce the degree of disturbance caused by frequent entry into the paddocks. The open area would serve not only the openness, but also little grazing land, as Ostriches like to graze during the day time. Stocking rates would vary from situation to situation.

Normally, under this system 3 acres of land would be required for one Ostrich and a group of 50 to 100 Ostriches can be reared in an area of 150 to 300 acres. Group mating would be followed in this system.

Advantage:

- Less cost involved in fencing and erecting boundaries
- Relatively easy to identify the non- performing bird
- Collection of eggs not difficult
- Individual birds can be identified for any ailment
- Feed cost can be substantially reduced, as the birds will have large areas for grazing

Disadvantage:

- Actual identification of male bird that has mated would be difficult to identify
- Problem would arise due to the presence of dominant male and female
- Handling of birds would be difficult in large paddocks
- Provision of feed into the paddock would be difficult

Semi-intensive system with artificial hatching

This system is almost similar to the previous system, except that the hatching of chicks would take place under artificial conditions. Since the hatching is being done using an artificial hatching system, in addition to the above features, chicks raising room is essential along with subsequent provisions for rearing the growers. After the grower stage, these birds can be accommodated in the breeding paddocks

Semi - intensive system of rearing

(Courtesy; PGRIAS, Kattupakkam, TANUVAS, Tamilnadu, India)

Intensive system

This system of rearing requires less space and more experience hand for regular farming activities. This is suitable for commercial ventures. The area required for the intensive farming system is normally less than 20 ha (50 acres), divided into small paddocks of 1-2 ha (3-5 acres) each. This system is popular for an Ostrich farming system, because of the small land area involved. However, there are two clear disadvantages associated with the intensive system. First, higher feed costs, as birds need to be provided with all their feed and, second, the higher cost of fencing. Capital investment is higher per unit of land area. Nevertheless, the advantages of using the intensive farming system are numerous and by far outweigh the disadvantages. An area of 0.3 acres per bird would be required. Small paddocks are being constructed to house the Ostriches either in 1:1 or 1:2 male-female ratios.

Advantage:

- Accurate and scientific record keeping for all the birds. These records are invaluable for assessing the eventual value of the stock, whether for resale, retention for breeding, or slaughter. Selective breeding of quality birds can be performed
- Egg collection and feed record keeping is possible. Further, accurate feed consumption records can be maintained, which enable operators to assess the feed conversion efficiency of the stock.
- Birds can be closely observed / monitored for any unwanted changes or behaviour
- Handling of bird is easier

Disadvantage:

- High capital investment in terms of housing and other facilities
- More expenditure on concentrate feed
- Incompatibility among the breeder birds would be noticed immediately and would require replacement

Among the various systems of management studied so far, it is wise to come to a conclusion that the extensive system of rearing Ostriches is not feasible due to high land cost and lack of sufficient knowledge on Ostrich farming. Semi-intensive system is possible after gaining some knowledge in Ostrich farming. Intensive system would be ideal for new entrants. This system would give them some hands-on experience on Ostrich farming. This system demands less area among all other systems of management, but would require more skilled labour and technical expertise. Nutrition and feeding would require more expenditure than expected. Specialized incubator would be required for successful hatching. Controlled breeding can be followed and regular culling would ensure good performing flock.

Intensive system of rearing

(Courtesy: PGRIAS, Kattupakkam, TANUVAS, Tamil Nadu, India)

Male and Female Ostriches

(Courtesy: PGRIAS, Kattupakkam, TANUVAS, Tamil Nadu, India)

Chapter

5

HOUSING AND PADDOCK MANAGEMENT

Management plays a key role in successful Ostrich farming. Rearing of chicks is one of the most important steps of the whole process. The age of 0-12 weeks (3 months) is the most important part of management, where the highest mortalities are experienced. As a developing industry, it will be imperative that through sharing experience, research and knowledge, we learn to reduce the loss by chick mortality. In fact, while calculating the cost of production, mortality especially chick mortality consumes bulk of cost. Many factors are involved in chick rearing and they are: genetics, health status, management, weather conditions, nutrition, brooding practices, facilities and environment, and stress conditions. Ostrich producers worldwide often experience high mortality in chicks and young juveniles as the result of improper brooding and poor early management practices. Caretakers must be trained and properly supervised to maintain desired conditions and to recognize and correct problems then and there. Ostriches are classified into chicks (0-3 months), growers (3-12 months) and adults (>12 months).

Stocking density and housing requirements

The stocking rate and minimum area required for Ostrich rearing is given below. This is taken as a basic guideline and the actual requirement may vary from farm to farm.

Age	Shelter	Paddock	Size
0-1 month	1.2 sq. m/bird	1.5 sq. m/bird	15
1-3 months	2.5 sq. m/bird	5 sq. m/bird	15
3-6 months	3 sq. m/bird	25 sq. m/bird	40
6-14 months	4 sq. m/bird	50 sq. m/bird	40
Pre-breeders	5 sq. m/bird	250 sq. m/bird	60
Breeders	5 sq. m/bird	1000 sq. m/trio	30

Brooder / chick facility

Chick house must be built such that it is convenient to maintain optimum temperature, humidity and ventilation and can be similar to any poultry brooder house. The chick house must be convenient to conserve heat and also to provide good ventilation. The chick pen must be rodent proof. All-in-all-out system of housing birds according to the age groups improves the bio-security. Provision of open area attached to chick house is preferable as it not only provides run space for young chicks, but also facilitates cleaning and quick drying of the pens. Avoid extreme climatic conditions (chill and hot) in the house.

The floor must not be smooth, preferably rough and solid cemented surface or hard soil is required which will be easy for cleaning and disinfection. Further, materials like stone powder or paddy husk or any other moisture absorbing material can be used as litter material, to be spread on the floor for better sanitation. Placement of burlap or gunny bags over the litter not only helps the chicks find better foot hold but, also facilitates easy removal of litter daily, thereby improving sanitation. Partitioning chick house into smaller pens of 15X15 feet avoids smothering and huddling. Partitioning of pens and open area facility can be done with wire mesh of 1 X 1 inch for better security.

Expected stress factor

Stress is one of the major causes in chick mortality. Chronic stress causes depression of the immune system and increased susceptibility to disease. Young chicks can die from stress alone without the presence of any symptoms / disease. There are many factors that can create a stressful situation for chicks, some of them seen very obvious and some are very subtle. Some of the common factors observed are;

- Poor ventilation
- Under or over use of medication
- Transportation
- Extreme environmental fluctuations
- Imbalanced nutrition
- Overcrowding
- Unnecessary handling

Grower house

As birds grow, they require larger space and also free run space. Provision of open run space will allow the chicks to go out for few hours during day time, thus exposing them to outside environment. A separate grower house with run space away from chick enclosures improves the bio-security. Chain link of mesh size by 2X2" is required for grower houses and the height must be 6 to 8 feet depending on the age of the birds. It is generally 1.5 times the height of the birds as these birds will attempt to jump and escape. The maximum height provision up to 10 feet must be made at adolescence age which they reach by 12 months. A gate size of 6 x6 feet is adequate. The floor should be of soil type. Keep away all foreign objects particularly larger stones.

Housing of a greater number of birds requires attention for cleanliness and sanitation. The removal of excreta as far as possible helps in keeping the pens clean. Open type feeders attract the wild bird and pests hence, special feeders for avoiding such birds and also to protect from direct exposure to sunlight and rain. For this, a specially designed feeder with a height of 4 feet for comfortable feeding by Ostrich should be provided. Water should be provided by bucket system at a height of 4 feet. Provision must be made to run off the rain water from all the pens without entering into other pens.

Ostrich breeder pens

These pens are required to house the birds for breeding purpose. It is clearly explained in Breeder Management chapter.

Design of Ostrich facilities

(i) Double-sided comb design

There are hundreds of possible designs to choose from. The most widely used is the simple rectangular design with a long service area (corridor) running down the centre of the pens as given below. Each pen has its own run extending outwards on both sides of the central service corridor. This is, appropriately, called the "double-sided comb design".

Attached to the service corridor is a feed storage area. The length of each run is not less than three times the length of the covered area, to ensure that the birds have plenty of exercise. This design is particularly suitable for birds from the age of three to 16 months (depending on the area of each run).

(ii) Central composite design

Another design that is being adopted by large producers is based on the "central composite design". Pens and their accompanying runs fan out from a central point. The inner circle is used as a service area with easy access to all pens. Variations on the full composite design (half or three-quarters) can be used, depending on available land area. The shape of the service area can also be changed to a hexagon (for six-paddock design) or octagon (for an eight-paddock design).

Bedding materials

A bare concrete floor is not practical for birds since it is hard, cold, retains moisture and germs, and is abrasive on chicks' feet. If rough, it is difficult to clean and if smooth, it becomes too slippery in the presence of any small amount of moisture or dirt. Similarly, a dirt (soil) floor is not practical since it is virtually impossible to clean, retains odours and can easily lead to ammonia build up. Dirt floors also become dangerously slippery when the soil is mixed with faeces and urine.

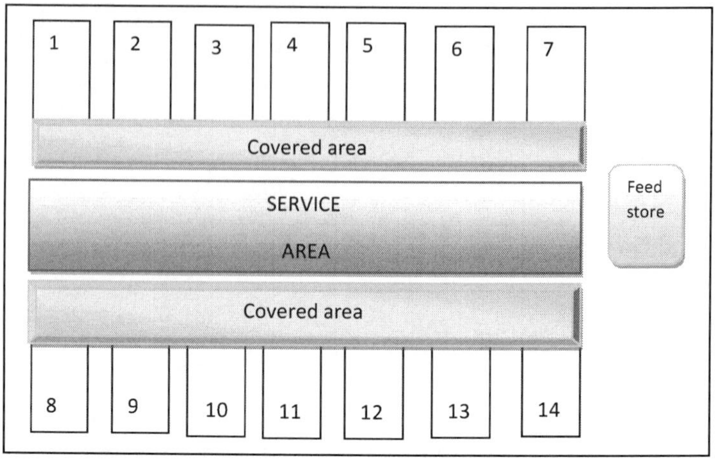

Double-sided comb design for Ostrich growers (above) and Central composite design (below)

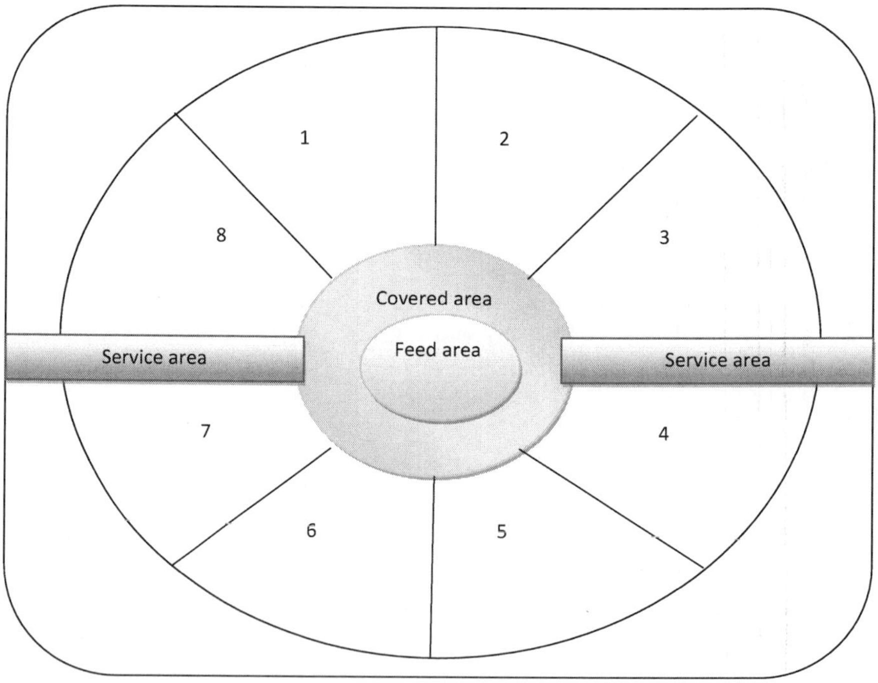

There are different types of "removable" bedding materials such as sand, sawdust (wood shavings), chopped straw, small gravel (pea gravel), etc. that have varying degrees of success and efficiency. Dust-extracted wood shavings are cheap, readily available and provide warmth for the chicks, in addition to having a high moisture-holding capacity. However, very young chicks may consume large quantities of the shavings and become impacted. Care should be taken to make sure that the chicks know where food is available. Removable floor materials such as artificial turf, carpets and rubber mats have also been used. High quality artificial turf and grass-like carpets are expensive and tediously difficult to clean. Moreover, chicks are often able to peck fibres from the surface, which accumulate inside the birds and cause impaction and death. Rubber mats are reasonably priced and appear to provide good flooring with ease of cleaning. However, the surface of the rubber sheets should not be too smooth.

Fencing

Fencing should be at least 150 cm high for birds up to 12 months and 165-170 cm for older birds. It must be easily seen by the birds strong enough to withstand birds, resilient enough not to injure the birds, free from projections and sharp parts and ensure the bird's head or legs cannot become trapped. The simplest form of fencing is five strands of high tensile galvanized smooth steel wire (3.1 mm gauge), mounted or suspended on round 15 cm diameter posts, 5 m apart. Between the posts, at least two "droppers" should be placed to act as visible upright barriers for the Ostriches. Corner posts should be round and 20 cm in diameter and there should be at least one midway post for extra strength. The first strand should be 25-30 cm above ground level, to enable the handler to make a quick escape if the situation arises.

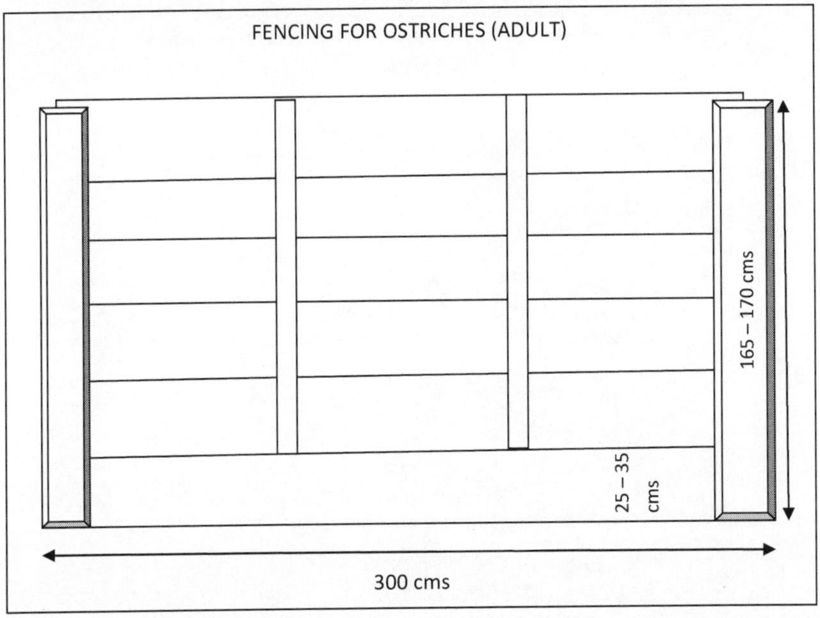

There are many variations on standard wire-strand fencing. A combination of woven wire (5 x 10 cm) similar to that used for sheep fencing can be used together with two strands of wire on 15 cm posts, 9-12 m apart. Sheep netting is used for the first 150 cm above the ground, with two strands above. This is particularly suitable for young birds. Rails or coloured builders' tape can be fixed between the posts to act as a visible barrier for the birds. A combination of 15 cm round posts and 5 cm steel tubes can also be used as fencing.

When the paddock is partitioned into two smaller areas of inhabiting the birds, the separated potion should be well marked with white tapes for the birds to see and halt. In the absence of such tapes, the growers tend to get injured by hitting the fence directly.

Rectangular paddocks for growers / adults

(Courtesy: PGRIAS, Kattupakkam, TANUVAS, Tamil Nadu, India)

Breeding trios in a paddock

(Courtesy: PGRIAS, Kattupakkam, TANUVAS, Tamil Nadu, India)

❑❑❑

Chapter

6

HANDLING AND RESTRAINING OF OSTRICH

Farmers have to handle Ostrich bird during various management procedures including their transportation and medication. Therefore, it is important that one gets familiar with their handling methods during various stages of operations. While handling the birds, it must be ensured that least stress and excitement is given to the birds and at the same time, the person handling the birds should not feel inconvenient. It is to be noted that animal welfare is also to be given due consideration while handling the birds. Ostrich birds are normally handled for the following purposes

- Sexing
- Care and management of sick or injured birds
- Moving birds from pen to pen
- Catching and loading for transportation
- Dividing a flock
- Separating fighting birds
- Physical examination
- Medication
- Show purpose
- Semen collection
- Artificial Insemination
- Banding and micro-chipping

Applying strength is not a major requirement in handling of Ostrich birds but adoption of proper and perfect technique is the need. Prior to handling them, it is important to beware of the strengths and weakness of birds. Normally, Ostriches are docile and cautious birds. However, when you attempt to work with them, their quickness and speed becomes very evident. While handling one must know very well about their strong powerful legs, sharp claws and quick feet. Ostrich always kick forward and hence, attempt to work from the side or from behind because of their ability to deliver powerful (200 psi) kicks to the front.

Usage of proper clothing, boots, gloves and long sleeve shirts are essential for protection from Ostrich kicks. It is always advisable to carry out various managemental activities during the cooler parts of the day, either early in the morning or late in the afternoon as this is the ideal time for the birds since, they will be calm and quiet during this period

1. Handling of Ostrich chicks (0 to 3 months)

Chicks of all ages, especially those from hatch to three months of age, should be handled with more care. At this age their fragile bodies are developing very fast and the stress brought on by rough handling can lead to death. The bones are likely to be not fully calcified, hence softer in nature.

While handling newly hatched Ostrich chicks, one can easily catch them by cupping their hands over the chick's sternum or breastbone to restrain its movement, then use the free hand to hold the chick's back, and then tuck its legs close to its abdomen. The chicks can also be restrained with a scooping motion from the front, with their hand under the sternum. Young chicks are always to be released by placing them on their sides, letting them get up by them. This will help to prevent leg injuries. Chicks of this size can easily be moved and restrained.

To pick up an Ostrich chick, place one hand at the base of the neck and one hand under the rear end, being careful not to squeeze or choked the chicks. To move an elder chick that is too large to carry, place yours hands in the same position and carefully restrain or guide the chick to its destination. Don't pick up the chicks by the legs or neck and ensure the chicks are settled gently.

2. Handling of Ostrich growers (3 to 12 months)

As Ostrich chicks grow, the task of handling becomes slightly more challenging. During this stage, the growing chicks are to be handled in a relatively smaller area with ease. If growing chicks are to be worked outside, utilize narrow fencing runs or alleys.

When handling birds at grower age, it is important to remain calm and be patient and try to minimize the stress on the birds. To catch a bird, it is best to corral it into a small holding pen. The holding pen must be sturdy and void of any gaps or holes in the fencing where the bird might get its head, toe or foot caught. This procedure not only expedites the catching process but, minimizes the chance of a bird running full speed into a fence and injuring itself. If, the birds are to restrain for longer period, then allowing the birds to huddle in fairly large area is recommended. The birds are then observed, this method is also followed during the counting of the birds.

3. Handling of adult Ostriches (> 12 months)

(i) Wing hold method

The most common way to handle adult sized birds is to approach the bird quietly (walk, don't run), once when we are close enough, grab it quickly and firmly from the side or behind by both wings. Allow the bird to get over its fright by letting it jump up or down a little while still keeping our hold on it. Then, if the purpose of catching the bird is to move to another location, propel it forward while moving behind or alongside it (still retaining the grip on the wings) so that the bird's energy is used for forward motion rather than jumping in an attempt to escape.

(ii) Stand-up technique/Two men wing walking technique

The stand-up technique can be used for quick banding, sexing, or parental administration of medicines, using two or three persons. The first person restrains the Ostrich by holding the wing and the other person may perform the task. After completion of work, the birds are released gently.

(iii) Using a Hood method

(a) Using a Hood – I

To move or restrain a juvenile Ostrich, a hood can be an invaluable tool. The hood is nothing more than a tubular piece of material with a large opening at one end for the head and a small elastic opening at the opposite end to fit snugly over a closed bill.

Once hooded, birds tend to relax and become much easier to work with. There are birds that will not require a hood and can easily be handled. However, to capture and handle the average bird, the following methods and techniques are recommended. If the bird is curious, somewhat friendly and routinely comes to the edge of the fence to peck at an outstretched hand, it is an excellent opportunity to hood and capture it.

Slip the hood over the handler wrist, large opening toward the person's fingers. When the bird pecks the fingers, quickly grab its beak with that hand. Once the handler has a firm hold on the bird's closed beak, use the other hand to slide the hood quickly over the bird's head, ensuring that it's vision is totally obstructed. Release the beak, allowing the head and neck to move freely.

Now, immediately move to either side of the bird and grab the wings. The bird can now be restrained and moved. When moving an Ostrich, another helper may assist by pushing from the rear, although this is not always necessary. In some case, a bird will be more willing to walk backwards. This procedure is fine. However, do not attempt to do this with young chicks, as their developing legs are extremely fragile and could easily be injured.

Restraining of Ostrich using a hood

(Courtesy: PGRIAS, Kattupakkam, TANUVAS, Tamil Nadu, India)

(b) Using a Hood – II

Depending on a bird's nature and comfort level around humans, the above method may not be feasible. Here's another method to try, if we have advance notice that a bird is to be captured. First, withhold feed for several hours prior to capture. When ready, put out feed as you usually do, and place the hood on your wrist as explained above.

When the bird goes to the feeder, it should be hungry enough to take feed directly from the palm of the handler hand. When the bird takes the feed, quickly grab its beak with the hooded hand and follow the instructions as mentioned above.

(c) Using a Hood – III

If these methods prove unsuccessful, enter the holding pen and slowly maneuver the bird into a corner. Once cornered, quickly reach for the back of the neck with one hand and gently pull its head down. Quickly grab the beak with the other hand and hood the bird. If a holding pen is not available and one shall be forced to capture a bird in a large open pen, first try to slowly maneuver the bird in to a corner and repeat the hooding procedure explained above.

(iv) Using a Hook

If the bird becomes overly excited and begins running uncontrollably into the fence, it may be necessary to utilize a hood designed specifically for catching Ostriches. They vary in length and are usually fabricated with aluminum. A hook can be dangerous to the bird and also to the handler if improperly used, so use it only as a last resort.

The successful and safe hooking of a speeding Ostrich requires quick and precise execution that comes only with experience. Never get in a hurry. If the bird becomes extremely excited allow some time to calm down. Then enter the pen and stand just beyond the path in which the bird is running. Firmly grasp the hood with both hands and hold it to the ground directly in the bird's path, with the open end of the hook facing the bird.

Restraining of Ostrich using a hook

(Courtesy: PGRIAS, Kattupakkam, TANUVAS, Tamil Nadu, India)

(v) Chemical Restraints

Drugs like Xylazine and Ketamine can also be used either single or both for restraining the birds. These drugs are likely to be used when there is a need to perform any surgery. This method can also be used just before transportation, as the birds would be still and the birds can be lifted and loaded. The dosage is very important, as higher dose can prove to be fatal. This has to be done under Veterinarian's supervision. The recovery part of this procedure should be well monitored as the recovery is violent and may be irreversible effect on the bird.

Chapter

7

SEX DETERMINATION IN OSTRICH

Sexing in Ostriches is not only essential but is of great economic importance. The breeders need to know about the sex of the chicks at an earliest time possible. This helps in identifying the males and helps in early culling or disposal of unwanted birds, especially the excess males. Accurate sexing is a matter of great importance to breeders. Many Ostrich breeders can attest to the time wasted in pairing birds of the same sex. In the majority of animals, a variety of phenotypic dimorphism is exhibited between males and females, but it is not always the case in these birds. Ostriches exhibit their sex differences as adults, so early sexing becomes relevant when breeders want to sell young sexed birds. Earlier the earliest period of sexual dimorphism was between 6 to 7 months, but with not 100% accuracy. Now with the advent of many latest technologies, sexing can be done at day one itself. Several methods for sexual determination in Ostriches are described below.

Molecular sexing (DNA typing)

DNA (Deoxyribonucleic acid) is the genetic "blueprint" that determines every characteristic of a living organism. Each Ostrich has its own unique DNA or nucleotide sequence, which can never be altered, so that even though Ostriches may look alike, no two birds will have an identical DNA. It is this that makes DNA analysis so valuable to breeders. In cells, DNA is located on structures called chromosomes. Sex chromosomes contain the genes that determine the sex of the individual. This test is 100% accurate.

The sex of all birds (including the Ostrich) is determined by the sex chromosomes. Male birds have two identical sex chromosomes called Z chromosomes (ZZ). Females have one Z chromosome and one W chromosome (ZW). When a male makes sperm, one of its Z chromosomes is carried into each sperm. When a female makes an ovum (egg), her contribution is different. If the Z chromosome is carried into the egg, the resulting embryo (ZZ) will develop into a male. If the W chromosome is carried into the egg, the resulting embryo (ZW) will develop into a female. The female chromosomes determine the sex of Ostrich offspring, contrary to the case of humans.

Surgical sexing

In this method an endoscope is inserted into the abdomen of the bird to see the gonads. Although the method has a very high degree of accuracy, it requires the use of anaesthesia and, therefore, carries some risk of anaesthetic death or injury. Furthermore, it is often difficult to sex young small birds surgically.

Vent sexing

This is the conventional sexing method that requires physical examination of the sex organs. The accuracy and safety of this method depend greatly upon the skill and experience of the operator. This may not give accurate results.

Male Ostriches have a black and white plumage when they reach an age of 7 to 8 month. Females and immature birds are much duller, with a greyish-brown plumage. The young have a spiky, black-tipped puffy plumage until they are about four months of age. The plumage of the cock is brighter during the breeding season. The skin, usually light blue, becomes bright red over the beak and forehead and around the eyes. The leg scales and toes become pink. The dusty gray-brown feathering of the mature Ostrich female or hen is thought to be of value in camouflage during the daytime incubation of the eggs.

Occasionally, mature Ostrich hens may exhibit a degree of black body feathering that varies from a few dull back feathers to nearly complete coverage of black feathers over the body. Colouring in Ostriches is influenced by sex hormones in that a lack of oestrogen produces black feathering in the male, whereas a high concentration of the hormone in the female causes the feathers to be grey or brown. Therefore, the rare black-feathered hen may not be as prolific as other hens because of low oestrogen levels.

The male Ostrich, sometimes called the cock or rooster, is a curious, rather docile but sometimes aggressive bird. Black body feathering and long white tail and wing feathers are characteristics of the male. This colouring of pigmented feathers is thought to serve as an effective camouflage because, in the natural environment, the male incubates the eggs at night and subsequently assists in raising the young chicks.

The male and female chicks are very alike and their sex can only be determined by examining their sexual organs. This is difficult as the penis of the male is still tiny and can be easily confused with the clitoris of the female. From about seven to eight months of age, sex can be determined when the bird urinates or defecates, since the penis emerges during these times. Unlike most birds, the male Ostrich has a penis, their urination and defecation are separate acts (although one normally follows immediately). Full distinction between the sexes is reached at about two years old. The wing quills are pure white in the male, while they are ringed with grey or black in the female. The tail feathers of the male are white or yellowish brown and those of the female are mottled light and dark grey.

It is important to note that the scarlet coloration of the male's beak and skin is dependent upon the presence of mature testes, while its black plumage is dependent upon the absence of ovaries (the oestrogen hormone in particular). A castrated cock never acquires scarlet coloration, but its feathers are the normal black of the male. Removal of the testes after sexual maturity has little effect on the bird's sexual instincts, and it will continue the mating procedure.

Chicks can be tattooed, branded, or micro-chipped at 1 day of age. The brand can be placed on the bare patch of the belly immediately behind the thigh. A thin wire can be formed into the identifying numerals or letters, heated red-hot and touched to the skin to create a permanent mark. Leg and wing bands also can be used for identification, but these may create health problems at a later time.

Chapter

8

CHICK AND GROWER MANAGEMENT

Successful rearing of chicks begins with purchase of quality breeders. The primary consideration should be sound genetics. Although the developments of genetics in chicken are far more advanced, the genetics selection in Ostriches is not as popular as in chicken. But, at least the history of breeder bird could be well ascertained.

Only well verified records should be used for consideration. Error at this stage may be very costly in terms of time and money. The mistake is irreversible and if attempted to correct, it may prove to be very painful. There are many components that will determine the health and survivability of chicks. It is extremely important that one should fully understand everything possible about Ostriches, hatching equipment, bio-security, genetics, breeder management, etc before planning to hatch Ostrich chicks.

Chick rearing

Chicks experiencing problems to stand (edematous chicks are more prone to this problem) due to weak legs or legs that splay out to the sides should be immediately hobbled for a day or so, using a half-inch strip of vet wrap.

- The affected chicks should be placed on its back to enable us to see the correct placement of the hobble.
- The strip should be wrapped around one ankle and it should be pressed to hold it in place. (Vet wrap adheres to itself.)
- The other leg should be pulled so that both legs are perpendicular to the body then the strip should be wrapped around other ankle and should be pressed in place.

The above procedure will help in reducing the problems that can develop from splaying. Vet wrap is a good product due to its softness on the chicks' puffy legs and its elasticity, which allows the chick to become mobile. The first 36 hours of hatching are very crucial to the chicks. Thermal regulation should be the primary object. The chicks should be allowed to bring its body temperature to 102oF and stabilize there for approximately 6 to 8 hours. Care should be taken that the body temperature of the chick does not go beyond 103.5oF. Either chilling or heat stress will lead to yolk sac retention.

After the chick's temperature gets stabilized, it should be placed in an enclosure where the temperature is controlled, allowing it a choice of temperature which ranges from 75 to 92 °F. The chicks should be allowed to make some choice, although this is a period where disorientation can lead to chilling or overheating if proper controls are not provided. Overhead heat and floor heat can be used. Simple articles can be very appropriate, such as a play pen with a heating pad. The conditions should be well monitored with well-placed thermometers. The chick should be weighed and a leg band is fixed for easy identification. All procedures performed during hatching, including assistance in hatching weight date medications given etc should be documented for future reference.

Tagging of one-week old chicks

(Courtesy: PGRIAS, Kattupakkam, TANUVAS, Tamil Nadu, India)

Combining the chicks

All the chicks can be combined in one single brood, if the age of the chicks is within three months. Usually the chicks accept new pen mates at a young age very readily. As long as adequate feeding stations are available for the chicks which are lower in the pecking order, there should not be any problem.

Chick enclosures

The thumb rule is, whenever possible the enclosure should be built in larger area. Chick with grow out facility should be at least one acre since the chicks grow out very fast. In a one-acre pen, 60-70 birds can be grown. Again, large is always better. An American Ostrich Association sponsored research project in Kenya observed that chicks housed in pens with 100 square feet per chick developed impaction, leg problems, pecking and other problems that are commonly observed. Chicks that had 250 square feet per chick from hatch to three months of age developed none of these problems. Stress is a major factor in crowded pens, especially at less than three months of age. The chicks always tolerate crowded grow-out facilities better than the older ones – at least until sexual maturity. The main reason is their immune system functioning which is much better after three months of age. The outdoor pens should be always cleaned of faeces every day in the age group of less than three months.

Chicks in the age group of one to two weeks old

(Courtesy: PGRIAS, Kattupakkam, TANUVAS, Tamil Nadu, India)

Chick condition

Signs of healthy chicks

- Enthusiastic chicks will run out of the night enclosure in the morning, twisting and twirling.
- They will exhibit curiousness about their surroundings – chasing moving objects, pecking at stones, sand through their beaks, pulling out loose plants, leaves, etc
- Except for short break in hot sun, the chicks are always seems to be busy all the time.
- Vigorous drinkers of water, they keep drinking lots of water to beat the heat of the day.
- General posture is head up, highly alert, with a full-bodied appearance – full neck and stout full legs with no veins or tendons showing.
- They tend to observe almost everything taking place around them

- There is very little one can do for such healthy chicks. Providing them clean water and nutrient enriched feed is more than sufficient.
- Their pens/ night shelter areas need to be cleaned on daily basis.

Signs of unhealthy chicks

- The chicks would not be happy to come out of the night shelter, the movements would be sluggish and looks "uninterested" with surrounding activities.
- No high posture, low neck carriage and hump-backed appearance.
- Skinny neck and legs
- Staring blankly and not keeping up with pen mates or standing and sleeping while others are busy.
- Any chick that stands out from the crowds is most likely to be verified for their health status.

A chick which has had surgery or injury is more prone to impaction. Since the bird is not feeling normal it will tend to eat all the wrong things. Post-op chicks have about a 75% higher chance of becoming impacted than normal chicks. Water soluble vitamins and electrolytes for poultry could be added (at the recommended level) to the drinking water for the first 1 to 3 weeks. The fat soluble vitamins - A, D^3, E, and K - are compounded with a starch or protein emulsifier to enhance dispersion and availability in water. This will ensure an adequate intake of vitamins, particularly A and D_3. Other additives are not needed. Depending on the season, if night temperature is above 40 °C or more, three-month-old chicks can stay out during night, provide the enclosure is well protected. A shelter to keep feed dry and to allow chicks the option of staying dry is a far-sighted measure. This is only humane for animals of any species.

One-week old chicks in brooding
(Courtesy: PGRIAS, Kattupakkam, TANUVAS, Tamil Nadu, India)

Ostrich producers often experience high mortality in chicks and young juveniles as the result of improper brooding and poor early management practices. The following recommendations, when effectively implemented and carefully followed, will significantly improve livability and quality of both chicks and juveniles.

- Caretakers must be trained and properly supervised to maintain desired conditions and to recognize and correct problems.
- The brooding facility must be designed to protect chicks from predators (dogs, foxes, etc.) as well as from inclement weather.
- Chicks should never be allowed to get soaking wet. It must be kept dry and under good sanitation at all times and should be designed for effective ventilation and ease in cleaning. Concrete floors in brooding units make them easy to clean.
- Temperature at chick level should be 88 to 92oF during the first 10 days of life, then 80 to 85oF until they are 3 weeks of age. From 3 through 8 weeks the ideal temperature is between 70 and 80oF.
- Chicks must never be exposed to chilling temperatures nor allowed to become overheated. Chicks brooded in small pens with raised wire floors and heated with infrared heat lamps are particularly susceptible to chilling in cool weather.
- Warm room brooding (uniform temperature throughout the room) will prevent such problems. Space heaters or central heating is recommended.

At 6 to 8 weeks of age, chicks can be ranged outside in good weather, but they must be sheltered at night. They can be managed in groups of 25 to 50 birds.

Day old chick in basket

(Courtesy: PGRIAS, Kattupakkam, TANUVAS, Tamil Nadu, India)

Grower rearing

Regardless of age, birds of same size should be raised together to minimize injuries. At one year of age chicks become more territorial and introduction of new chick or juvenile should be avoided to an already established group, otherwise it provokes fights. Placing new birds in an adjoining pen with a common fence line for a week or two is a way of minimizing these problems. The feed from chick to grower should be changed from six months onwards, with grazing is preferable. This is one of the reasons for successful beginning. The birds can balance their own diets of given a choice.

If Ostriches are allowed to forage on natural grasses or grain, the nutrient intake should be balanced with commercial feed. Ostriches should not be allowed to become overweight. Excess body weight in the form of fat is detrimental to egg production during breeding and to meat quality of birds that are to be slaughtered.

4 Month old bird

(Courtesy: PGRIAS, Kattupakkam, TANUVAS, Tamil Nadu, India)

The earlier stage of grazing should be done very carefully, as the growers are very likely to swallow to anything and everything. Pens must be well drained, clean, and free of any coarse, dry vegetation, pebbles, small rocks, wire, staples, and other debris which may cause intestinal obstruction or death, if consumed. The feeding part of the birds should be well taken care of. Spoiled and mouldy feed are highly detrimental to Ostriches. After 6 months of age, the growers are fairly hardy and can be ranged outside with less danger of intestinal obstruction. Shelter and shade must be available to protect both birds and feed during inclement weather and at night.

One-year old grower birds

(Courtesy: PGRIAS, Kattupakkam, TANUVAS, Tamil Nadu, India)

Shifting to new environment

Chicks may be shifted to the grower or larger paddocks after a period of six months. This is another crucial phase during which heavy mortality can be expected. As soon as the growers are being exposed to the new environment, they will tend to eat indiscriminately which leads to impaction of digestive tract, which is fatal in majority of cases. The area may be totally fenced with at least 7 to 8 ft of border fence. This protects the birds from being straying away and also gives protection to the workers. During this time, the birds tend to become more aggressive. The change over from one feed to another should also be gradual, possibly at least over a period of seven days. It is always advisable to identify the sex of the chicks as early as possible and rear them separately. This would help in avoiding infighting among the growers and also helps the managers to plan their future activity i.e. to retain them as future stock or to be sold for meat. Generally, the birds suffer from stress quite easily and this can be easily observed in Ostriches. Birds exhibit stress by fluffing out their wings, hyper ventilating and keeping their beak wide open. As far as possible, the birds should be separated at an early age and housed. Separation of birds at a later stage causes stress among the birds. During loading and shifting of juveniles, sufficient area should be given for movement of the birds and for the persons who are involved. Separate entry and exit should be provided for the movement of male and female Ostriches.

Transportation of juveniles is necessary for movement of birds within the farm or while shifting from one area to another or for slaughter / marketing. Inside the farm, the birds can be blindfolded by the use of hoods and shifted without any disturbances.

10-month-old birds

(Courtesy: PGRIAS, Kattupakkam, TANUVAS, Tamil Nadu, India)

Chapter

9

BREEDER MANAGEMENT

Ostriches are seasonal breeders and the breeding takes place only during particular seasons of the year. Their breeding or mating season lasts for six to eight months each year. However, the timing and duration of breeding vary with latitude and altitude. In the northern hemisphere, breeding commences in March and ends around August/September. In the southern hemisphere, it commences around July/August and finishes by the end of March. Due to global change, recent reports indicate that the breeding activity also gets either delayed or in some cases it has also been advanced by at least a month or more.

Selection of Breeder

Selection of chick need not be based on the performance of parent stock alone. That is to say that heredity does play a very large role in future performance. However, we are dealing with a bird that reproduces for longer period. In its reproductive life time under intensive management of at least 40 years, we can expect it to produce at least 432 chicks (@12 livable chicks per year). In no way, every off-spring of a top-quality pair of birds will also be top quality. When choosing future breeders for a farm, one must look at the overall survivability of the chicks that hatched, the number of chicks hatched versus the number of fertile eggs laid by the hen, and also the degree of difficulty involved in raising the chicks. Conformation is another much overlooked when choosing future breeders. A well-formed bird should be somewhat light in the chest and should carry itself in an upright position when walking in an alert manner. Bird's natural tendency to slump forward when relaxed should not be taken as criteria. However, the body carriage is determined by the overall conformation of the bird. A bird should have more body mass from the centre of the thigh to the tail than from the centre of the thigh to the front of the chest. This allows the bird to stand more upright, and have a downward slope from the base of the neck to the tail. This is important in excreting body fluids which otherwise could be retained, allowing for a higher possibility of reproductive tract infections.

Birds should have a gigantic appearance; a bird with a long neck should also have long legs. Conditioning is another area that should be looked at closely. The appearance of the feathers is very important. They should be clean and fluffy, not brittle and matted. Of course, this can be some-what deceiving after a rain, when most birds look very rough. As desert animals, the Ostrich's feathers are not capable of repelling water. The bird's spinal cord should not see externally. The bird's feathers can be pulled to check for the stress lines. These are long, very distinct lines running through the feather that can indicate stress and or illness. If the eyes are the mirror of the soul, they can also be a mirror into how a bird feels. The eyes of the birds should be clear and alert. Legs should be straight. Some birds will have their toes pointed out at a slight angle, but provided the angle is not severe and is equal on both feet, it is acceptable.

The overall look of the bird should be smooth sleek lines and an easy way of moving. A defect such as a crossed beak, crooked neck, or rolled, broken, crooked neck, or rolled, broken, or bent toes, is not acceptable.

Adult male bird – pre breeding stage
(Courtesy: PGRIAS, Kattupakkam, TANUVAS, Tamil Nadu, India)

Breeding behaviour of females

The wild Ostrich is sexually mature at four to five years old, while the domesticated Ostrich is mature at two to three years; the female is mature slightly earlier than the male. Some domestic Ostriches may start their first breeding season (producing fertile eggs) as early as 18 months of age. For the most part female Ostriches are very docile by nature. Though most do not want to be touched, they can be handled without much problem. With proper nutrition and a good environment, about 50% of two-year-old hens will lay eggs. Many of these hens will lay only in small numbers as two-year-olds, but others will lay prolifically. Once again we are dealing with individuals where genetics, environment and nutrition play a great part in overall results.

Females in heat (ready to breed) will drop their wings all the way forward, often touching the ground, and shake them, some-times making the joints sound as though they are popping. This action is referred to as fluttering. Usually at the same time a hen flutters she will drop her head to the ground and clap her beak. This is called clucking. The entire procedure is called displaying, commonly referred to as "courtship display" in hens. This type of behaviour is brought about by the high levels of oestrogen present in the hen's body, which is a sign that laying should follow within 30 days or so. Hens that flutter for 60 to 90 days and do not lay an egg should be checked by a veterinarian to make sure that there are no problems.

In rare cases, fluttering behaviour can be seen in female chicks at three months and up. This does not mean that the chicks are ready to breed, but rather that they are checking their bodily responses. When seen in young chicks this behaviour is only momentary. It isn't until a hen begins to flutter and cluck continuously that she is generally ready to accept a male for breeding.

When the young Ostrich hen reaches maturity she will usually begin to flutter at the beginning of longer days. This posturing activity is an indication of the sexual receptivity toward the male. In this sexual display the hen lowers her head and neck to body height, snaps her beak, spreads her wings and begins a rhythmical wing trembling or fluttering. This sexual posturing activity is sometimes called clucking but this is perhaps a poorly chosen term, for in other avian species this term may indicate broodiness. This activity becomes very frequent at the beginning of breeding season and indicates her continued receptiveness toward the male.

Breeding behaviour of males

The flashy white wing and tail feathers play a prominent role in the courtship ritual or sexual display (Kantling) of the Ostrich rooster. The neck and head feathers are short, sparse and usually gray to dull white. Red-pigmented skin develops rapidly on the beck, neck and lower legs at the beginning of the breeding season. These skin and attitude changes are hormonally induced and are influenced by light, season and environmental changes.

The regal and rhythmical movements become more aggressive during the breeding season and the male will defend the nest, eggs and female very vigorously. Matured male Ostriches may also have a distinct enlargement around the vent, especially during the mating season, due to the enlargement of the phallus in its natural position on the floor of the proctodeum. A young Ostrich rooster may mate with a female (fake mating) by two years of age, but may prove incapable of producing viable sperm. Most males produce fertile sperm by three years of age; however, in their natural environment they may not be sexually mature until four years of age.

Male bird making sound during breeding season
(Courtesy: PGRIAS, Kattupakkam, TANUVAS, Tamil Nadu, India)

Anyone who has had any dealings with animals in breeding situations is well aware of the fact the males of the species are more aggressive during the breeding season. The Ostrich is in no way different. The male Ostrich will warn the onlookers by flipping it's wings over his back and hissing. During these times, direct contact with Ostriches must be avoided. The construction facilities must be such that, one should be able to collect eggs without antagonizing him. The breeding male is no more dangerous than a stallion or bull during breeding season, if common sense is used.

Ostriches have a variety of vocalizations, but are normally silent. During the breeding season, territorial males produce a deep booming trisyllabic "ohh-oooh-ooo", descending in pitch at the end. To attract females or against other males, the call is uttered mainly by day (especially early morning) but occasionally at night and can be audible up to 3 km away. Other calls include a soft "booh" or "twoo", which are possibly contact calls and may also be associated with anxiety and distraction displays. In addition, hoarse guttural calls, "hurp", hisses, beak snapping and stomach rumbling are produced in threat or aggression. Chicks in anxiety or distress utter entirely different, melodious, trilling "quirrn-quirmr" calls.

Adult male

(Courtesy: PGRIAS, Kattupakkam, TANUVAS, Tamil Nadu, India)

The mating session can occur for at least 45 to 90 seconds. The first mating of the day will last longer than the last mating of the day. Both male and female are promiscuous, with each seeking multiple partners. The male will mount the female from back and often the condition of the feather on the left side is an indicati0n of mating. The male will rest the left leg on the left side of the female hen. After the first mating, the male will start searching for secluded place for the female hen to lay the eggs. The hens usually lay the eggs in the late afternoon or early morning. Both male and female birds, sometimes display clucking behaviour next to the nest; it is usually associated with the presence of eggs in the nest or the male bird's solicitation towards the female to produce an egg.

Adult female

(Courtesy: PGRIAS, Kattupakkam, TANUVAS, Tamil Nadu, India)

Breeding dynamics

(i) Pair (1+1)

A pair of Ostriches is just that: one male and one female making up a pair. For the most part pairing breeders is not a problem. If we are introducing birds that were not raised together, it is a good idea to put both into a new pen at the same time. Moving, pairing, or any other types of changes during the breeding season should be avoided as far as possible. The breeding pairs, if new, should be placed in their breeding pens at least sixty days prior to the beginning of breeding season. New breeding females should never be introduced in the middle of the season.

(ii) Trios (1+2)

Trios consist of one male and two females, which allows more production and less pen space. This is preferable, if long term breeding is planned. Most males can handle breeding more than one female. If we have a pair and he is breeding the hen five or six times or more a day then we should think about making a trio for the next breeding season. New breeding females should never be introduced in the middle of the season. More than likely one or the other or both will try to kill her and could succeed if not observed properly. The breeding pairs/ trios will exhibit territorial behaviour and would seldom allow new bird.

To introduce a new hen to an established pair of breeders, the wise thing to do is to place the hen from the pair and the new hen into a new pen together and sufficient time to get acquainted should be given. This must be done before the onset of breeding season. Usually after a couple of weeks they will hang around together, eat together and often sleep together. The pen will become the hen's territory and they will set the rules. Now we can introduce the male to the pen. This method is much safer and will have fewer problems. In the trios, also there are chances that the male may avoid one female totally. In that case, the female may be shifted and aligned with another male for the next season. Aggressive male will usually serve both the females, if alignment is accepted by all three.

Red coloration on the legs of male Ostrich indicates the onset of breeding season

(iii) Colony mating (1+ 2 to 3)

Colony mating has also been practiced widely, where the number of birds is huge, few hundreds to few thousands. The fertility is expected to be much higher than the previous two systems, but the possibility of death due to fighting among the birds is also higher. The biggest disadvantage is pedigree which cannot be fully adopted in this system. Further, it is also difficult to identify the hens which lay the eggs. This system can be a success, if all the birds have been reared together at similar / same environment. A genuine breeder would avoid this system.

Facilities required for breeding

Pen location

The ideal location for breeder pens is in a very secluded area away from traffic and the everyday workings of a ranch. These birds love trees and shrubs. If possible, at least one large tree in any part of the pen may be given and yet plenty of space is allowed to soak in the sun during the winter months. If a tree is out of the question in the area, then an artificial shade of some type, preferably an extended lean-to in front of their shelter, facing the south may be provided. When building the shelter facilities, the wind directions should be taken into consideration. The pen should always be lengthwise unlike the growers where, circular type is preferable.

Pen Size

The size of pen should be minimum 5000 sq ft per bird. Hence, for a trio an area of 15000, sq ft may be provided (equivalent to 0.34 acre) Males and females do not always come into season at the same time. If possible all corners should be 45-degree angles, to prevent the male from cornering a hen and / or hurting her. A 14 to 20 foot wide alleyway between breeder pens can be allowed, which can double as an extra pen for the purpose of holding an injured or sick bird, or just to separate the birds in times of any emergency. Closed alleyways also give the birds a safe haven against free-roaming predators, such as stray dogs. The birds learn that the predators can't get close to them when they are against the alley fence. As a general rule they will run to this safety zone and will not run fences, which can lead to injury or even death.

Breeding Paddock

Generally, larger the paddock better will be the health. However, since a large area is undoubtedly more costly, common sense should be exercised when deciding on how large or how small the breeding paddock should be. The Royal Society for the Protection of Cruelty to Animals (RSPCA) of the United Kingdom recommends no more than 12 adult birds per acre (0.4 ha). This figure represents the upper limit and is not necessarily the optimum number. Ostriches are lovers of space and prefer large breeding areas.

Clearly, farmed Ostriches cannot be provided with very large areas (unless land prices are low), but they should have sufficient space in which to forage, walk, run and display courtship when they wish. Accordingly, a minimum of 0.1 ha (0.25 acre) paddock is needed for two to three adult Ostriches in breeding (one male and two females). If larger numbers are to be kept in one paddock, then the paddock size should be increased exponentially (i.e. a minimum of 0.3 ha/0.75 acre for four to six birds, 0.6 ha/1.5 acres for 8-12 birds, and so on). Ostriches are likely to breed better if provided with more than the minimum space.

Adult laying hen

(Courtesy: PGRIAS, Kattupakkam, TANUVAS, Tamil Nadu, India)

Shelters/Holding Pen

Shelters for birds need not be fancy or expensive. A simple two – or three- sided structure, approximately 12′X 12′ to 16′ X 20′ will be suitable for three adult birds. The birds should be given feed inside the shed only. The newly purchased birds some time may refuse to go into the shed for feeding. They can be slowly trained to take the feed, by placing the feed in a tub outside of the shed or where they will eat. The feed pan can be moved two to four feet per day toward the inside of the shed.

A small enclosure may be facilitated around the shelter, which can serve as a holding room during treatment, cleaning the shed, collection of egg, etc.

Fencing

- There are several types of fencing material that are suitable for Ostriches. The best is 9-gauge chain link (ordered with knuckle top and bottom) using 2.375″ pipe for posts and top rail. However, this type of fencing is very expensive.

- The next choice would be 2X4 no-climb horse wire. This type of fence does not require a top rail. However, if birds hit this wire they will be more apt to lose feathers and cut themselves than with the chain link.
- The next type most recommended would be tight-lock game fencing, eight feet high. This wire will give if hit by the birds but due to the height does not seem to cause problems. The more room we give the birds the fewer fence problems we will have, and more room allows for less – expensive fencing.

Photoperiod

When birds are exposed to artificial long days during the winter there is a prompt regeneration -of the testes and ovaries, imitating the changes that occur naturally. Conversely, if birds are kept on short days, little or no gonadal development occurs. As for as birds are concerned, long days are stimulatory and short days are inhibitory, while neutral days are those containing 12 hours of light (naturally occurring along the equator). It is important to note that the stimulus must be maintained if the response is to continue. It is well known fact that light affects reproduction mainly in two ways. Light or photoperiod determines the onset or termination of the breeding season. Light also acts like a biological clock, which determines the time of egg lay on a given day. There is always a debate in the affect of light on reproduction especially in the birds. The species where selective breeding has been practiced for ages, the light bring sexual maturity much earlier than the species where selective breeding is not in vogue.

Eggs are usually infertile during the early part of the breeding season. This is usually caused by infertility in the cock. Breeding cocks should receive 16 hours of light per day beginning 3 to 4 weeks before being penned with their hens. After being placed together, hens and cocks should receive 16 hours of light per day during the entire breeding season. If the Ostrich farm is located in an area of the country which has a natural day length longer than 16 hours, get the time clock to match the longest natural day length. After natural sunset, 2-5 foot candles of artificial light intensity at head height should be adequate to stimulate and maintain egg production. Three of four days of diminished light or shortened day lengths may cause the entire flock to almost completely stop egg production. Although there is not much work done on the effect of photoperiod on sexual maturity in Ostrich, there is every reason to suspect that Ostriches cannot be anyway exception to other poultry species. Currently, their activities are correlated with the natural day light. Moreover, Ostriches are also referred to as opportunistic breeders. In addition to the existing day length, other factors like environmental temperatures, availability of drinking water, vegetation, etc kicks off the onset of breeding season in birds. An early rainfall can trigger early breeding season due to availability of lush (pasture) green vegetation. A long dry summer can extend the non – breeding season further. Hence, the season of breeding can never be a fixed regime.

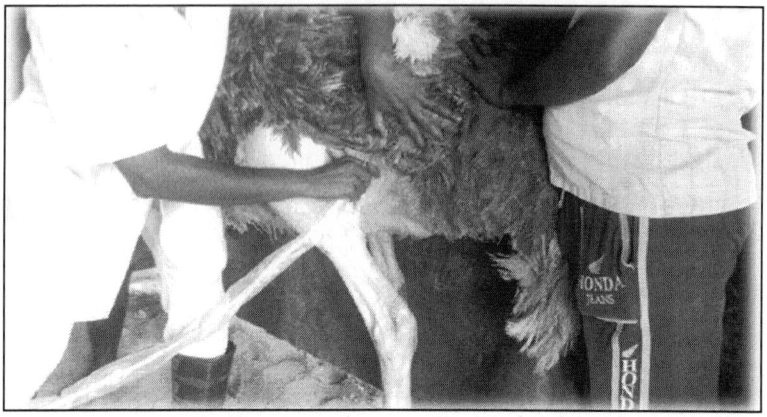

Ultrasonography in adult female to check the egg follicles
(Courtesy: PGRIAS, Kattupakkam, TANUVAS, Tamil Nadu, India)

Egg Laying

The female starts to lay fertile eggs shortly after mating. The first fertilized egg is laid approximately 10 to 14 days after first mating. Thereafter, and almost without exception, eggs are laid every other day in clutches (sequences) of 10 to 14 eggs. The hen stops laying for a period of seven to ten days after which she starts a new clutch again. 60 to 90 eggs are laid during the breeding season by high-producing females and some hen produces less than 40 eggs per season. The highest number of eggs recorded is 105 eggs during one season. The lowest has been 25 eggs per bird per season. Similar trends were not observed in subsequent laying seasons. In any given breeding season, the females "come into season" before the males so that early eggs are very likely to be infertile. The first laid eggs are in the weight group of 600 to 800 gms. The later eggs have been recorded to be an average of 1300 to 1600 gms. The heaviest egg recorded was 2150 grams. It is usually advisable to allow the breeding pairs from the second season onwards for the females. Though the males are sexually matured at two years of age, the males should be allowed for breeding based on the size of the organ.

However, at this point in this industry it is not well established whether the hens that lay large numbers will continue this pattern for life or whether this could cut their laying life-span in any way. In general, a good strong hen that will lay 50 eggs per year with a good hatchability and survivability rate in her chicks is far more profitable than one that lays a large number of eggs where hatchability and chick survival is a question.

The eggs are laid in a communal nest on the ground. The first female to lay in the nest is normally the one that subsequently undertakes the guarding and incubating, and is termed the "major hen". Any bird that lays in another female's nest is called a "minor hen" because of the minor role she plays. While minor hens may lay their eggs in several nests, a major hen will only do so in one nest. Compared with the minor hen, the major hen will spend a relatively longer time at the nest, and will attend to the nest daily even on the days she lays herself. When a minor female arrives, the major female "gets up" usually within a few minutes, and waits 5-20 m away from the nest until the minor female has laid and left. During the actual incubating period, the major hen will remove the eggs of minor hens and would incubate the eggs laid by itself.

Other unusual things that may be encountered are:

- A hen may lay an egg with no shell near the beginning of her cycle, or
- She may lay an egg with bloody streaks on it, or
- She may lay an egg standing up and drop it to the ground, breaking it.
- On most occasions, the bird will always sit and lay the eggs and would turn around to see the laid eggs.

Before they set into a real cycle, many start to lay before they have actually been bred, some will strain very hard, others will sit or stand many times before their eggs are laid. Most of these problems are simply due to immaturity and will correct themselves within a week or so. The point to be concerned is when a hen consistently lays eggs with no shell or wrinkled shells, or if she does not establish a correct laying pattern of every other day. If this pattern continues for two weeks or more, remedial measures have to be sought.

Eggs are normally laid in a shallow scrape. They should be collected twice daily. Reproductively active cocks can be extremely protective and aggressive. Caution should be taken when entering their breeding areas. Aggressive cocks can be fed and penned in a catch or holding pen while eggs are being collected. If the eggs are not removed, the female will start incubating them during the day, leaving the male to incubate from dusk to dawn. The male's jet-black feathers assist in hiding the eggs during the night. For commercial Ostrich farming, it is important that all eggs are removed from the nest at least twice daily, since, if allowed to incubate, the female will stop laying until the chicks have reached four to five weeks of age, resulting in financial loss.

Recent developments in breeding of Ostriches

Genetic evaluations of all farm animals are being done to improve the efficiency and productivity of the animals. Ostrich is no exception to such developments. Pedigree and production records need to be maintained to use the genetic tools effectively. The available data are processed by Best Linear Unbiased Prediction (BLUP) methods. This software would separate the influence of environment and genetics. This is one of the best genetic tool currently available to estimate the genetic worth (breeding worth) of an animal. Since the life span and reproductive age of Ostriches is more than 30 years, use of this tool would be effective to assess the "value" of our available stock. The breeding value of the stock would help us in estimating the performance.

Last two decades have seen tremendous development in the genetic evaluation of farm animals, especially dairy cows, beef cattle, small ruminants, poultry, etc. The animals which are reared under intensive system are being subjected to these tools to increase their production potential economically. Some preliminary studies have been carried out in Ostriches also. Similar use of such tools in Ostriches is also feasible when the data of individual birds are available. This is mostly possible in breeding flock and not directly applicable in commercial flocks. This involves huge investment in terms of infrastructure for breeding birds. Accurate data collection and analysis is the key to success.

Some progress has been made to study the performance of South African Black (SAB) and Zimbabwean Blue (ZB) cross breeding and using the genetic tools to study their performance as individual genotypes and all cross genotypes. The hybrid vigour is estimated to be around 6.7%. Use of a third genotype (Kenyan Redneck) is under investigation for the breeding plan. The details of the breeding plan and use of genetic tools have been documented at the Directorate of Animal Science, Elsenburg Research Farm, Western Cape, Department of Agriculture.

Reproductive Performances as observed under Indian conditions at the Ostrich Breeding Unit, PGRIAS, TANUVAS, Kattupakkam, Tamil Nadu, India

Ostrich egg

Ostrich birds lay one of the largest eggs (1300-1500g). Eggs of Ostrich represents 1.2% of its body weight but still it is the least when compared to chicken (3.5%) and quails (8%). The egg colour varies from white to yellowish white and the surface from rough to smooth. The length of the egg is from 16 – 18 cm and width 13 – 15 cm.

Laying season in Ostriches

It has been often informed as the Ostriches are seasonal layers and their breeding season also varies from place to place. It is quite true as many references are suggestive of the same. In India almost similar situation has been observed, the breeding season is quite true, but the laying is observed round the year and the margin of difference is minimal.

Age at sexual maturity and egg numbers

- mean age at sexual maturity : 810.76 ± 46.16 days
- minimum age : 593 days
- maximum age : 1398 days
- maximum no. of eggs in a year/ female : 62
- minimum no. of eggs in a year/ female : 5

Pattern of egg laying in Ostriches

Months of laying	Percentage
Jan – Mar (I Quarter)	(32.2%)
April – June (II Quarter)	(20.3%)
July – Sep (III Quarter)	(22.5%)
Oct – Dec (IV Quarter)	(25 %)

Adult male incubating and hatching the eggs
(Courtesy: PGRIAS, Kattupakkam, TANUVAS, Tamil Nadu, India)

❑❑❑

Chapter

10

MATING BEHAVIOR AND FERTILITY

Since ages, Ostriches have been inculcated to reproduce and hatch chicks during that period of year when sufficient food and water are available for sustainability and survivability of their offspring. The birds in different parts of the world have different breeding seasons. The birds breed when the total photoperiod is comparatively higher than other months. In some place the birds breed immediately after the rainy season, when the vegetation is rich enough to sustain the chicks.

Physiology of male and female reproductive system

The copulation organ of male is called as phallus and it lies folded inside the proctodeum compartment of the cloaca. The average length of the phallus differs from age from 25 cm to 45 cm. The size also varies depending upon on season or off season of breeding. The size is expected to be bigger during main breeding season. During the actual time of mating, the male will service from the left to the middle portion. The main reason is the phallus is curves to the left when it is out of the compartment.

The hen Ostrich has single ovary on the left side only. The ovary will be flat grey disc during the non breeding season. And it will lack any follicular activity. The reason is that during non breeding season, there will not be enough circulating reproductive hormone like FSH or oestrogen. During the breeding season, the ovary will assume like bunch of grapes. The ability to lay eggs depends upon the environment, nutrition and the presence of male birds. Under natural conditions, the clutch size varies from 8 to 14 eggs in a nest. As per the recorded observation, it can be said that more than one hen can lay eggs in a single nest and a single hen to lay in more than one nest. Under such conditions, it is impossible to record the performance of the birds.

Four phases of the mating act are distinguished:

1. Initiation of mating (Courtship / display behavior)
2. Mounting
3. Intromission and ejaculation
4. Post-coital display.

Initiation of mating

The male Ostrich is a majestic creature. During the non-breeding season he is quite docile for the most part, willing to take a bite of grass from your hand and in general allow you to come into his territory without much fuss. Of course a lot of this depends on how much time you spent with him since chick stage. Males will begin to get their black feather plumage somewhere between the ages of eleven and eighteen months, depending on the maturity of the individual. Some males can mate result in fertilise eggs prior to two years of age, and some at three to four years of age. However, they are considered prolific breeders. The breeding male is more aggressive with typical territorial behaviour during the breeding season. The change in colour of the un-feathered areas (beak, vent, shins, etc) from pale pink to deep red is sure indication of onset of breeding season.

Kantling

The male's sexual display will consist of both "booming" and "kantling". Kantling is a very interesting procedure whereby he gets down on his hocks (first joint of the leg above the ankle), and extends his wings while swaying back and forth and touching the ground. It is used for two purposes, one to ward off intruders from his territory and second to entice his female companions. This will make us believe that the male is in "happy mood". A male will sometimes sway from side to side as many as sixty or seventy times in one sitting. And the whole process will last not more than 100 to 120 seconds. Sometimes this process is repeated twice or thrice at a stretch.

Clucking

Female Ostrich will exhibit broody behaviour indicating her readiness to be mated. This will be characterised by female hen walking with her wings stretched side wards and shaking them gently. During this time, her head will be close to the ground and pecking aimlessly at the objects. This is also referred as "clucking".

Mounting

The male after the sexual display will go near the females by thumping its foot repeatedly in short strokes. The female will understand the situation and will go down on it's knees (hock joint). This will encourage the male to sit on the left side and mount the female hen. At this juncture the male phallus will be at the maximum size and deep red in colour.

Mating in Ostriches

(Courtesy: PGRIAS, Kattupakkam, TANUVAS, Tamil Nadu, India)

Intromission and ejaculation

The male will mount by placing one foot on the hen's back and the other foot on the ground near her side and insert his phallus while striking his head on his sides and swaying back and forth. The female holds her wings horizontally and flutters the tips. Her head is held low, and she opens and shuts her beak. When she is crouching, the male places his left foot next to hers and mounts with his right foot on her back. His penis passes into the cloaca of the female. Consummation is usually accompanied by male groaning wildly, while the female snaps her beak and shakes her head. This completes the process of mating.

Post-coital display

If copulation is achieved, the female will move her head back and forth with her neck extended and both birds will then vibrate their wings and make a loud moaning sound. As with most bird species of the world their mating ritual is unique and fascinating.

Factors affecting reproductive efficacy of the breeders

1. Environment factors

Apart from the overall environmental factors, like rainfall, temperature, etc, the bird's actual environment like breeding paddock, status of male, time given for acquaintances, age of the breeding birds, number of available birds in same paddocks, etc also plays a major role in the breeding performance. Female birds are comparatively very sensitive than the male birds to the change in partner during breeding. The frequency of egg collection also influences the mating frequencies among the birds. Early removal of eggs will prolong the mating season among the birds. Location of breeding paddock plays a significant role in determining the mating of the birds. Location nearer to public movement, periphery of the farm, public roads, highways, will hamper effective mating in the birds. Any form of disturbances would result in stress and the result will be reflected in the mating and ultimately the reproduction in Ostriches.

2. Bird related factors

The differences between the achieving the puberty and sexual maturity should be well understood as in case of mammals. Puberty can be achieved before the actual sexual maturity can be achieved. Partners should be formed based on the sexual maturity. Puberty is age when the sperms are formed in males and ovarian follicles are formed in females, but not good enough to cause fertilisation. However, in sexual maturity both the genders are fully developed, functional and the levels of respective reproductive hormones are sufficient enough to result in fertilisation. Age at which the sexual maturity occurs differ in different breeds of Ostriches. South African Black (SAB) neck reach sexual maturity at three years of age, where as the Kenyan Red neck and Zimbabwe Blue neck reaches a year later at four years. The breeding birds should have good libido to sustain the breeding performance. If the breeding among the birds is reduced due to libido, it is very difficult to identify the cause and may take years to return to normalcy.

Fertility

Fertility in male birds should be given more importance than in female Ostriches. Fertility in male is determined by genetics, environment, nutrition, stress, etc. Frequent semen evaluation, maintenance of good breeding records, egg production, hatching performances, etc will be true reflection of the fertility. The defect in female birds is not uncommon. Semen evaluation will be true indication of semen viability and health of breeding males. Female birds can be subjected to ultrasound testing to look for the presence of ovaries during the breeding season. The breeder birds should receive high plane of nutrition with adequate nutrients required for the breeding. Any imbalance found in any one ingredient will pull down the total effect of the nutrients. The nutrition of the breeding birds during breeding is slightly different from breeding birds during non breeding season. The rest period is also equally important to replenish the lost nutrients during breeding times. As far genetics is concerned, the reproductive traits have low heritability which makes it very difficult to achieve genetic progress in due course. The disadvantage in achieving the desired genetic progress is the long generation interval.

Feeding of adult male Ostrich

(Courtesy: PGRIAS, Kattupakkam, TANUVAS, Tamil Nadu, India)

❑❑❑

Chapter

11

ARTIFICIAL INSEMINATION IN OSTRICH

Artificial Insemination (A.I) is a part of assisted reproduction technique which has been used in cattle, sheep, goat and poultry has received considerable attention in Ostrich industry also. An acceptable protocol can be formulated for AI in Ostriches, which will have wider and deeper impact in the industry. The profitable production of Ostriches requires increased numbers of offspring, in numbers exceeding those to be found among birds living in their natural habitat. The problem of unfertilized eggs has been identified as one of the most important limiting factors for the success of breeding programmes. Establishment of all female flock should be the ultimate achievement in producing the chicks profitably.

As the reproductive performance of the male Ostrich exerts considerable influence on the efficacy of the fertilization procedure, the assessment of sperm quality with regard to spermatozoa motility, morphology and density could be valuable instrument for evaluating the fertility level of male Ostriches. Male birds from the age of 4 to 6 years will be used for training, collection and evaluation. The males should be kept in small groups separated from female Ostriches.

The criteria for selection were:

1. Good reproductive records (fertility above 70% and egg production of over 40 eggs per year). All reproduction records should be analysed combining both farm and hatchery results.
2. Sexual imprinting of both the male and the female to humans: all males displayed their specific courtship behavior (dancing head shaking) in order to induce female to crouch. When females crouch the male will conduct full sexual intercourse. Some males will start courting when visualizing humans, especially people who are visiting the birds on continuous basis. Courting the human will stimulate the female to crouch. In addition, some females can also be stimulated at the presence of humans, as manifested by wing waving, beak

clicking and crouching. Therefore, when humans enter the reproductive area a sexual cascade could be conducted by the birds.

3. By the beginning of the commercial reproductive season, birds should be trained utilizing a constant routine of auditory and visual stimulation. These stimuli could be included in daily visits by the researcher to the reproduction yard and vocalization toward the mating pair. One month later, at the end of the training period, birds were mating in the presence of the researcher, who was able to safety enter the reproduction yard and collect semen during natural mating.

(a) Semen collection techniques

1. Collection of semen in an artificial vagina could be designed as per the local need. The artificial vagina could be made of double sided with a latex condom of a size to fit the Ostrich phallus (60 cm length, 15 cm diameter). The end of the condom was connected to a plastic test tube. When the mating procedure starts, prior to penetration, the trainer places the male phallus into the artificial vagina until the ejaculation is completed. The artificial vagina was then firmly shaken to let the semen down into the test tube.

2. Collection of semen after the termination of natural mating by using a funnel could also be tried. In this system, the Ostriches are allowed to mate. After mating, the male would rise, when the phallus is still out of the cloaca a funnel (15 cm. in diameter) can be placed underneath to collect the semen drippings directly into a test tube attached to the funnel bottom.

3. Collection of semen using a tygon tube connected to a low vacuum device (turkey semen collector). Just before the start of the mating, the person would enter the reproduction paddock, thus enabling the male to mate with the female. After verifying that the male had started, the person would place himself under the male's tail and insert the tygon tube into the semen canal of the phallus. Semen will be aspirated into a glass jar under low vacuum pump pressure. At the termination of natural mating the person gently pulls the tygon tube out.

4. Collection of semen using a glass test tube. As in 3 above, except the person would insert a test tube into the semen canal, allowing semen to drip into it by gravity.

5. In this method, the harmony between stimulating dummy to stimulate males and the use of artificial vagina to collect the semen from the stimulated males should be well established. The male birds should be trained many times to adopt this method of collection. The next step is, the dummy is kept into the pen through a special hole in front of pen to stimulate the male. When the male have the peak of sexual stimulation the artificial vagina with long arm can be put into the special safety hole. The semen collection will be done from out of pen by another person, because of the fierceness of males in the mating season. The warm water is used to keep artificial vagina the same temperature

of Ostrich body and the vagina moved into the male penis and entered from a suitable distance. After many movements, the male will ejaculate the semen into the artificial vagina and it is collected in a sterilized plastic gradient tube connected with the artificial vagina.

6. Another method is, the Ostrich could be placed in a plucking crush and as of bag could be pulled over its head to keep the bird calm. An assistant could insert his hand into the cloaca in order to remove the phallus. It should be then held in an extruded position using a soft and clean pad for a firmer grip. Some Ostriches could ejaculate without further stimulation; some others could require soft massage of the papillae seminales, which is located by inserting two fingers into the bird's proctodeum. Semen samples to be collected in sterile glass tubes.

(b) Semen evaluation

A small portion of each sample to be used to prepare smears for microscopic assessment. Ejaculate samples should be collected throughout the entire breeding season, as well as in the non- breeding period in order to evaluate the seasonal effects on semen quality parameters. The volume of the ejaculate is measured in a 2 ml glass cylinder; the colour and consistency will be assessed by visual inspection. The concentration of hydrogen ions in each sample to be measured using pH paper and ranged from 6.4to 8.0. The degree of mass motility to be evaluated using light field illumination microscopy of ejaculate samples on pre warmed glass slides at a magnification of 45 X. In order to evaluate the level of forward motility, a cover slip is applied to a drop of semen and the mean percentage of linear motile sperm estimated using a phase contrast microscope at a magnification of 45 X. The examination of Ostrich ejaculate samples was completed on the basis of principles already established for other domestic poultry species. The macroscopic and microscopic investigation techniques and the staining methods proved to be adaptable without modifications.

The ejaculate volume varies greatly, measuring within arrange from 0.1 to 1.5 ml with a mean value of 0.64 ml. Normal colour ranged from white to ivory and the physiological consistency ranged from thin to thick creamy. Some ejaculate samples taken at the beginning of the season showed deviations in colour samples displaying a brownish colour indicated contamination with blood and should be excluded from the evaluation. The pH ranged from 6.4 to 8.0, with an average value of 7.3. Sperm concentration counts ranged from 3.9 to36.0 million/µl (mean value: 7.3 million/µl). A mass motility with distinct wave-like swarming movements was found in 65% of the ejaculate samples. Little movement in 20%of the ejaculate samples was noted, and in 15% no mass motility was visible. Individual sperm motility ranged from 42% to 96% in the sperm samples, with a mean value of 78% for forward-moving individual sperm. Regarding the morphology of the sperm, 5% to 26% may be abnormal (mean,17%) and 4 to 28% (mean, 20%) may be dead.

Table 2. Values of selected essential parameters in Ostrich spermio- grams

Characteristic	Unit	Mean ± SD	Range
Volume	ml	0.64 ± 0.22	0.1 - 1.50
pH		7.3 ± 0.1	6.4 - 8.0
Concentration	Million/μl	16.4 ± 5.2	8.9 - 78.1
Individual motility	%	78 ± 4.2	42 - 96
Abnormal sperm	%	17 ± 3.7	5 - 26
Dead sperm	%	20 ± 5.1	4 - 28

Peak sperm concentrations were found at the beginning of the breeding season; the lowest values were found at the end of the breeding season. The highest quality ejaculate was obtained from those males whose semen was collected only once a week. The basic study provided fundamental data for the establishment of minimum quality requirements for Ostrich sperm to be met by individuals receiving certification as breeding animals and for the selection of suitable males for use in artificial insemination.

(c) Artificial Insemination

Artificial Insemination technique is useful in overcoming the fertility and hatching in Ostrich to increase its rate of production. In the beginning, the pens of Ostrich should be designed for Ostrich artificial insemination and the males are isolated from females with special doors to let males indoor when mating.

The methods appear to be reliable, although there is still room for improvement. Female Ostriches develop a responsive behaviour and crouch sexually, thus allowing insemination to be carried out without restraining the birds until recently, female Ostriches needed to be caught and restrained for this procedure, which requires up to four people. A behavioural approach, as developed for female birds, has also been used for female Ostriches but responsive females had to be identified and accustomed to this procedure. The semen can be deposited into the vagina despite anatomical differences in their cloacae.

Artificial insemination technique

1. The first one trained females was used and the collected semen was artificially inseminated into female reproductive tract at the female's sitting posture, sterilized modified pipette with rubber end to protect the reproductive duct from any injury which affects directly on egg production in the future.
2. The second technique was used to fertilizing the non-trained females which fertilized in standing position in special pens, same modified pipette has been used to protect the female reproductive duct from injury and bleeding.
3. The direct semen collection technique from the female reproductive duct after normal mating will have significant results in semen quality, fertility and

hatching percentage as compared with artificial vagina technique with male stimulating by the dummy. Hatching percentage amounted 40-50% in normal mating before using these techniques. The AI results of recent times showed the ability to use these new techniques.

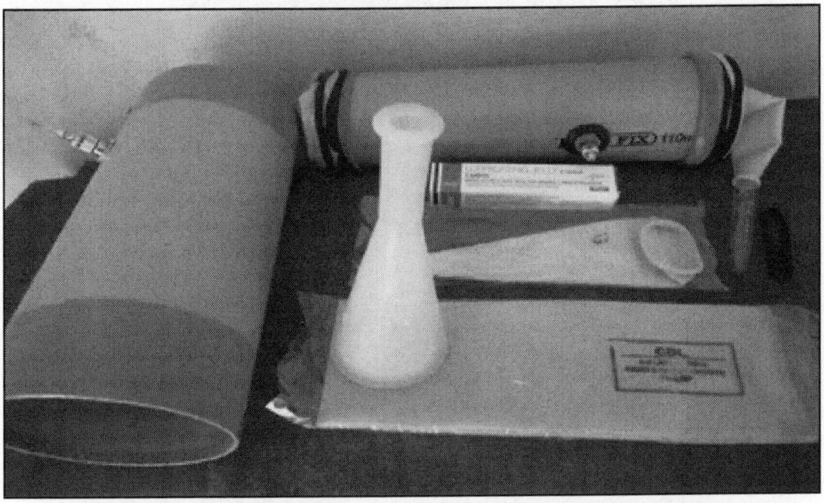

Designing and preparation of Artificial vagina for semen collection
(Courtesy: PGRIAS, Kattupakkam, TANUVAS, Tamil Nadu, India)

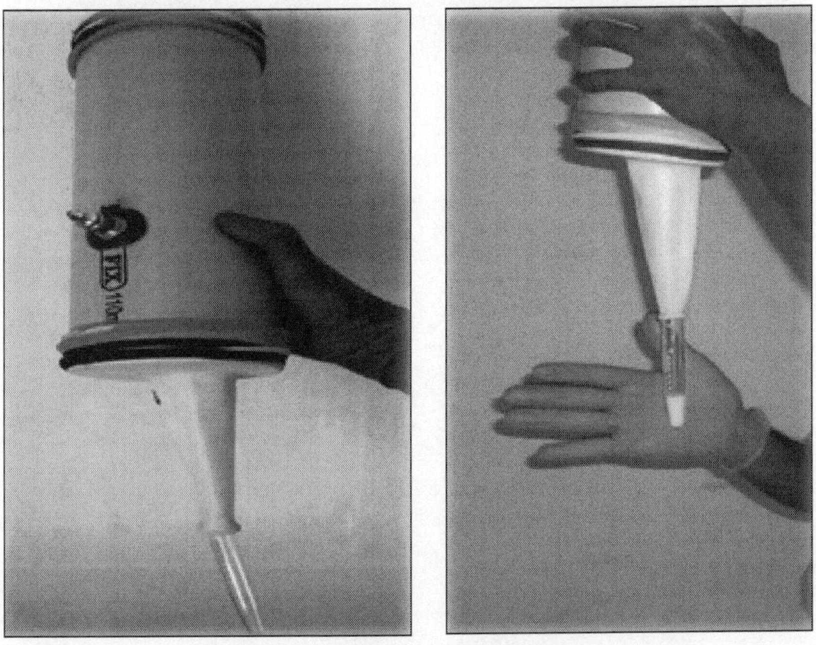

Before and after collection of semen
(Courtesy:PGRIAS, Kattupakkam, TANUVAS, Tamilnadu, India)

Collection of semen from male Ostrich bird
(Courtesy: PGRIAS, Kattupakkam, TANUVAS, Tamil Nadu, India)

Artificial insemination in female Ostrich
(Courtesy: PGRIAS, Kattupakkam, TANUVAS, Tamil Nadu, India)

Chapter

12

GENERAL BEHAVIORS AND THEIR INDICATIONS

The Ostrich has been bred for more than one century, but its behaviour has attracted little attention. Most of the citation dealings were of mostly nutrition, management and veterinary importance, while only few are involving behaviour. Hence, comprehensive information has been given for the benefit of the Ostrich growers. Observation of Ostrich for their behavior provides useful information about the activities being enacted by the birds.

General behaviour

Normal behaviors; Normal behaviors pertain to activities of daily living, such as, communication, socialization, grooming, feeding, and sexual activities. They are normally expressed similarly by all members of a species.

Activity-related behaviors

Twirling; Twirling is a behavior in which an Ostrich will twirl around, often until dizzy. This behavior is most often noticed in the morning when birds are let out of their indoor enclosures. Sick birds rarely twirl.

Thermoregulatory; In order to dissipate heat, birds will open-mouth breathe and open their wings. They have no sweat glands. Conversely, to keep warm, birds will often bend down covering their legs with their wings.

Pecking; Pecking is a normal exploratory behavior in Ostriches and is a predominant behavior in birds along with sitting, walking, eating, and drinking. It is, however, mainly a discriminate behavior with Ostriches because they attempt to eat almost anything.

Grooming; Ostriches lack an uropygial glandthus feather preening is not a predominant behavior. Dust or sand baths are the preferred method of grooming in these birds.

Trembling; Trembling is a behavior that resembles shivering. It is often noticed in birds suffering from stress, such as when cornered or transported. Thus, it can be a manifestation of anxiety and should not automatically be interpreted as an indication of hypothermia.

Social behaviors

Threat; In a typical threatening posture, the bird will stand tall with tail erect, hissing, wings slightly open, and feathers puffed up, especially around the neck. This behavior is usually noticed in juvenile and adult birds of both sexes, although most often exhibited by males.

Kick; A forward leg kick is most often performed by juvenile and adult birds, especially during the breeding season. It is an outward sign of aggression.

Vocalization; Two distinct sounds can be heard in an Ostrich flock: distress calls and booming. A distress call is most often emitted by chicks in distress such as, when they become separated from the flock or transported. Booming is a sound produced by mature males that serves to establish territorial boundaries and warn other birds to stay away.

Submission; An Ostrich exhibiting submission will run away from an aggressive bird or drop to the ground without defending itself. An overly submissive bird paired with an aggressive one may die from stress, injury, and exhaustion.

Sexual behaviors

Clucking and fluttering; A breeding hen may express her physiological readiness to breed by emitting a clucking sound made by rapidly opening and closing her beak. Simultaneously, she may flutter her wings by dropping them low and forward, and vibrating them in sequence.

Kantling; This is a typical male territorial behavior in which the bird drops to his hocks, and fans both wings forward and backward while hitting his head on each side of his spine.

Male bird in Kantling pose

(Courtesy: PGRIAS, Kattupakkam, TANUVAS, Tamil Nadu, India)

Soliciting; Female bird shows this display as a part of her willingness to mate where she hold her wings forward and down flapping them backward and forward while holding her head close to the ground opening her beak repeatedly to make a clapping sound.

Agonistic display; It is the sign of aggressiveness exhibited by male towards neighboring male/female birds or other animals.

Courtship display; Importance of this behaviour is that it leads to copulation which is often initiated by monotonous booming sound (male). Reports proved that unsuccessful attempts at copulation are mainly due to mating without Kantling display. Courtship display diminishes in frequency in both male and female when the Clutch is being laid as the bird usually goes for brooding

Abnormal behaviors

Feather-picking; This is a behavior in which a bird will aggressively peck feathers from the back and tail area of pen mates. It is often brought about by stress, overcrowding, boredom, and frequently follows a strong seasonal pattern in juvenile and adult birds, being most severe in winter months because of prolonged confinement. Once a bird has lost significant amounts of plumage, it is usually best to put a blanket on its back to allow for feather re-growth and provide protection from the cold. In young chicks, a genetic component has been suggested for this behavior, as some hatchlings will start feather pecking when only a few days old.

Toe and face pecking; Excessive toe and face pecking can lead to mutilating wounds where entire eyelids are pecked out. This aberrant behavior is most notorious in young chicks. Although causes are unknown, it has been suggested that stress, overcrowding, and boredom are predisposing factors. It is imperative that aggressive chicks be removed from pens, as this behavior can become contagious.

Behavioral stargazing; Aberrant behavior in which bird will continually lift its head up and back to the extent that it eventually touches its spine. The behavior is directly related to confinement. Severely affected birds will have difficulty walking, eating, and drinking, because of the abnormal position of the head. A diagnosis can be quickly con- firmed by putting affected birds outside. If confinement stress is the cause, the head will be straightened up almost immediately. This type of stargazing should be differentiated from the one induced by thiamine deficiency reported in poultry. This behavior occurs commonly in inadequate facilities where pens are too small and too dark.

Anorexia and adipsia; Cases from several farms have been documented where entire pens of birds will go off feed or water. Birds can develop aversions to either feed or water containers, or the location of such containers, to the point of ignoring them completely. If these birds are not supplied with an alternative, they will starve or dehydrate them- selves to death. Similarly, they can develop an aversion to feed or water, such as, when the feed is moldy or the water contaminated with high chlorine levels. Again, even after correction of the initial problem, birds may still die of anorexia or adipsia if not given alternate feed or water containers, because the aversion to the previous containers is so great.

Dietary indiscretion; The commonest health problems and the most life-threatening behavior in Ostriches, especially younger ones. Ingestion of foreign material is prominent in birds of all ages and is initiated by stress or curiosity. This behavior can lead to impaction, gastric stasis, and perforation of the digestive tract by sharp objects. Material commonly ingested includes whole alfalfa, sawdust and straw used as bedding, hardware used in pen and fence construction, rocks, grit, and sand. To prevent occurrence of gastrointestinal problems from dietary indiscretion, it is best to choose the bedding material with care to keep stress levels to a minimum. Sweeping pens with a metal detector on a regular basis is a good idea to prevent hardware disease.

Pica; The ingestion of feces seems to be a normal behavior in young Ostriches. Hatchlings in the wild will actively search out adult feces, presumably to colonize their gut and thus facilitate absorption of food. However, in captive situations, this behavior may become excessive, especially if birds are raised on a solid substrate, such as cement, and are not provided with environ- mental stimulation. Excessive fecal ingestion can lead to the accumulation of fecal material inside the beak, thereby interfering with the normal uptake of food. This fecal material seems to predispose chicks to respiratory problems, as they are inhaling air through this contaminated mass. To prevent excessive pica, feces should be removed from pens on a regular basis, and chicks provided with toys or appropriate fibrous material in small amounts to peck at.

Aggression; Aggression can be manifested toward other Ostriches or people. This behavior, in which one bird will intensely pursue and attempt to kick another bird, is most frequently observed with the onset of puberty. Aggressive birds should not be allowed to continue this behavior and should be separated from their victims. Pen modifications can be made to reduce the likelihood of birds injuring one another. This can include rounding of fence corners to prevent a bird from being trapped, addition of large round bales of hay in the outdoor enclosure as visual barriers, and physical separation of birds for parts of the day. Aggression towards people is more common in breeding males. They will display threat by puffing their feathers, slightly opening the wings, and erecting the tail. Further aggressive behavior will ensue by pushing the breast bone against a person and attempting a forward kick with one of their legs. With experience, one rapidly learns to interpret signs of aggression and position one- self safely in pens.

Aggressive behavior of male ostrich
(Courtesy: PGRIAS, Kattupakkam, TANUVAS, Tamil Nadu, India)

Chapter

13

INCUBATION AND HATCHERY MANAGEMENT

In comparison to poultry artificial hatching, the fertility in Ostriches to be around 70 ± 5 %, hatchability at 45 ± 5 %, early embryonic mortality at 25 ±5%, and dead in shell to be 25 ± 10%. Since the heritability is also low, desired progress as in chicken cannot be expected in Ostriches. However, best managemental practices can have the results as mentioned above. To get the best results, during breeding season, their nutrition requirements should be well taken care during the rest phase.

Natural Incubation

Natural incubation in Ostriches is shared between the male and the major female. The male incubates the eggs overnight, which represents about two-thirds of the 24 hours. The major female takes over from the male about two to three hours after dawn and leaves the nest one to two hours before dusk. The nest is seldom left unattended. The changeover is usually rapid and both birds are rarely present together at the nest. The male loses the bright red neck flush, distinctive of breeding males, halfway through the incubation period. Furthermore, the male will rarely interact with the female. Young males sometimes refuse to incubate but, after a few days of irregular sitting they take up their task seriously. The central temperature of naturally incubated eggs is maintained at 34-36 °C, with a nest humidity of around 42 percent. These levels are made possible by the use of a temperature recording device inside a fibreglass dummy egg. Natural incubation lasts for 42 days, with hatching success reaching a maximum of 100 percent, for central eggs, in many nests. The male will often become impatient and crack the egg himself, shaking it until the chick falls out.

Artificial Incubation

Since the advent of artificial incubation and hatching of Ostrich eggs many years ago, and together with a better understanding of incubation in the wild, much progress has been made in clarifying the circumstances that contribute to the success (or failure) of fertile eggs to yield young chicks. The incubation period of Ostrich eggs is 42-45 days and the actual process of hatching a chick is not at all simple. The environment in which eggs are incubated plays an important part, coupled with the position and turning of the eggs.

A satisfactory hatch of healthy chicks depends on the correct combination of breeding stock, their management, standard feeding and health status and proper handling and incubation of eggs. In order to get clean and dry hatching eggs, the breeders must be provided with a clean, dry place to lay the eggs. Eggs are usually laid either late in early morning or in the evening and majority of the birds will be settling to lay at the same time of the day every three or four days. Once the birds have settled into a regular laying pattern, it can be easily known when the birds are due for laying eggs so, that eggs can be collected.

Eggs should be picked up shortly after they are laid within 20 minutes after laying. Eggs should be collected with clean dry hands possibly with the use of disposable rubber gloves. Use of paper towels or wash cloth separately for collecting each egg is also recommended. It is suggested that the egg can be maintained in the same relative position as it was laid throughout the collection, cleaning and incubation process. In other words, don't turn it end over end or upside down while handling. While the eggs are collected it should be conveniently marked to identify the parentage or pedigree. A simple method is to apply a small strip of adhesive tape over which the pen number or hen number can be marked. Ifthere are only a few breeding stock a bucket with soft materials placed on the bottom can be used for collection of hatching eggs.

Egg collection

A perfectly hatch able egg may be ruined before it is even set, by bacterial contamination. Bacteria can pass through the pores of the eggshell within three to four hours of the egg being laid. Entry is greatly speeded up, if the shell is wet and dirty. When the egg is first laid, it has the Ostrich hen's body temperature but cools rapidly. This cooling causes the egg content to shrink and decrease slightly in volume. Since, the shell does not contract as much, a vacuum is generated in the egg and air and bacteria are drawn in through the pores, forming what is known as the air cell or sac. Bacterial contamination of hatching eggs can be greatly reduced by adopting the following guidelines.

Collect hatching eggs at least twice daily and more frequently at high temperature can be done.

- Use clean / separate containers for each egg.
- Clean the dirty eggs as soon as possible after they have been laid.
- Avoid wiping the eggs with a dirty damp cloth since, this is the fastest way to contaminate them. Use dry fine sandpaper for cleaning off large lumps of dirt.
- When washing the eggs, follow the manufacturer's instructions explicitly with regard to the concentration of either the detergent disinfectant or formaldehyde.

- Ultra-violet lighting (in the range 200-300 nm) is effective as a germicidal technique.
- Cool the eggs gradually before packing for storage.
- Avoid water condensation on eggs when traying.

The Ostrich egg

As befits the largest bird, the Ostrich lays the largest egg of any living bird. Oddly enough, however, the Ostrich egg is one of the smallest in relation to the size of the bird. Measuring, on average, 17-19 cm in length and 14-15 cm in width and weighing from 1500 g to 1900 g, the Ostrich egg is only just over 1 percent of the female's body weight. The eggs vary from white to yellowish white; their hard shiny surface is pitted with superficial pores of various sizes and shapes.

Hatchery building and equipments

Proper management is critical for successful hatchability. Hatchability problems can be caused by inadequate breeder nutrition, mating problems, improper egg handling, incubator malfunctions, and humidity or temperature problems. Successful management of a moderate size hatchery requires a high degree of expertise and attention. Hygienic conditions and sanitation is of paramount importance. Environmentally controlled hatchery building should be designed for durability and ease of cleaning. It should be of sufficient size to handle anticipated egg volume and must include areas for egg cleaning egg trays, cooling and storage, incubation and hatching, chick holding, equipment washing, and storage, as well as office and sanitary facilities. Equipment requirements for the hatchery include a stand by generator, forced draft setters and hatchers, service tables, a vacuum cleaner, pressure washer, tray washers and carts. Incubators for ratite birds are manufactured and sold by several commercial companies. A good ventilation system is a must for successful incubation and hatching. In addition, while built a hatchery, keep in mind that Ostrich eggs require a low relative humidity (24to35percent).Thus, in areas of the state with high relative humidity during the Ostrich breeding season, a dehumidifier system should be installed.

Egg washing

Egg washing with disinfectants will be preferred by few hatchery personnel's, after fumigation because, they feel it is easier, faster and also the residual disinfectant material remains on the shell surface and protects against bacterial contamination. Egg washing effectively sanitizes the egg, if properly carried out. Washing can contaminate if water temperature drops below recommended levels or contamination exceeds the capacity of the disinfectant in the immersion water. Wash water must be always hotter and the recommended range is 110-120°F (43-49°C). Several disinfectants are safe for hatching egg disinfection. Some of them are : A solution containing 250 ppm of Quartenary ammonium and 10 ppm of EDTA can be used and other disinfectants which have been recommended are Iodine (200 ppm),Chlorine –di-oxide(100 pm) and Chlorhexidine (200 ppm). Manure or dirt should be gently scraped off or lightly sanded with a fine grit sandpaper

Fumigation

This is used to control specific disease problems particularly omphalitis and salmonella infection. Among the disinfectants, Ionizing radiation and Aldehyde disinfectants only kill the micro organism as well as bacterial spores. That's only formaldehyde fumigation of eggs prior to storage is advisable before setting in Incubator. Eggs should always be cleaned and fumigated. Fumigation of hatching eggs prior to holding recommends 3 X concentration i.e. 60 g potassium permanganate crystals and 120 ml of 40 percent commercial grade formalin for every 100 cubic feet space. For proper fumigation the temperature and humidity should be 20°C and 65% respectively. Pour the measured quantity of the formalin over the potassium permanganate and close the inlet. The formaldehyde gas produced will kill the micro organism on the shell surface. After 20 minutes the eggs should be removed and graded for storage. The formaldehyde gas is poisonous and extreme care must be taken during the fumigation process. The maximum concentration of the formaldehyde in atmospheric air should not exceed 5 ppm. Then ventilate the area for at-least twelve hours before use. A brown colour residual powder is the good indication of proper fumigation, if it appears to be purple, then the quantity of potassium permanganate might be over dose, or if we find residual formalin liquid in the lid, it is due to over dose of formalin. In both the case, it is clearly indicate improper fumigation. A good cleaning with chlorine bleach and water followed by fumigation has proven to be very effective in killing all bacteria which may be present in the equipment.

It is extremely important that the incubator and the hatcher shall be disinfected prior to use each season with 4 x concentration. Formaldehyde solution is highly toxic and irritating. Always follow the directions on the label. Wear mask, head gear, goggles and liquid proof gloves while handling fumigants. Once the egg has been cleaned and disinfected, the process of marking and recording can be completed. Affix the tape firmly to the egg. Non-toxic grease pencil may also be used. Weight of the egg should be done before incubation. This is needed to identify the parentage and also hatchability and fertility of the individuals and to offer corrective measures, if any required.

Storage of hatching eggs and position

Ostrich eggs should be stored with the large end up. During this holding time, the egg should be rotated twice a day. Hatching eggs should be stored in a cool and if possible in a constant temperature room. The ideal storage temperature for egg storage is between 50 and 60°F (10 and 16°C). Better hatchability may occur if eggs are set (placed in the incubator) within 2 to 4 days after laying. Longer storage may reduce hatchability. Humidity during holding period is not as critical as temperature but, higher humidity prevents rapid evaporation. As a practical measure the humidity should not be so high that the eggs sweat when they are removed from the storage room. Moisture on the shell can be sucked in, taking with it the bacteria if any on the surface. Longer storage will reduce hatchability. Dirty eggs should never be set. Use of individual towel for each egg is preferable and the towels should not be reused.

Storage of eggs in cold room

(Courtesy: PGRIAS, Kattupakkam, TANUVAS, Tamil Nadu, India)

Transport of hatching eggs

Hatching eggs are to be frequently transported to hatcheries from the breeder farms. It may be transported by train, truck or other modes of transport. It should be always remembered that precautionary measures during holding period in the egg storage also applies to transport. Drastic changes in temperature should be avoided. They can tolerate gentle movements, but sudden jerks or jarring can cause considerable reductions in hatchability. Egg breakage can be avoided by using foam filler flats.

Position of eggs in the incubator

After proper marking the eggs are loaded in the setter trays vertically and they should be positioned to turn from end to end. Eggs loaded in the incubator trays should be placed with the air space end up.

Temperature and relative humidity

Maintenance of an optimum temperature is of prime importance for satisfactory hatching results. Embryonic development is initiated at temperatures far below the optimum, but only as the temperature approaches the correct level does hatching proceed.

Under natural incubation conditions with Ostrich parents and in still-air incubators (with heat applied from above), where a temperature gradient exists with the maximum temperature at the top of the eggs, the optimum temperature at the upper level of the eggs is between 38 and 38.5 C. In forced-draft incubators, with uniform heat distribution around the eggs, the optimum temperature is evidently near the centre of the potential hatching zone, between 35.9 and 36.5 C. These are the very narrow limits of temperature between which the embryo will develop properly. Towards the end of the incubation period, when the chick begins to produce its own heat, the temperature can be lowered by 0.7 C

Arranged eggs in setter

(Courtesy: PGRIAS, Kattupakkam, TANUVAS, Tamil Nadu, India)

Incubation time to hatch for Ostrich chicks is between 39 and 44 days with 42 days being about the average. The optimum incubation conditions for the Ostrich are in the range of 97.0 to 98.4°F dry bulb temperature with a relative humidity in the 25 to 40% range (wet bulb 67 to 73°F). The higher the dry bulb temperature the lower relative humidity needs to be. For instance, at a dry bulb temperature of 98.3 °F it is estimated that the relative humidity needs to be as low as 20% (wet bulb 68°F), whereas at an incubation temperature of 97.5°F the relative humidity may need to be as much as 30% (wet bulb 72°F). It should be remembered that the higher the incubation dry bulb temperature the more water that is produced in the egg, thus creating the need for a lower incubation humidity to remove that water from the eggs. Individual hatchery incubation requirements will be dependent upon the lowest wet bulb temperature that can be maintained in the setter room when the air conditioners are set on 100% fresh air. This will determine the lowest operating humidity of the incubators.

Turning eggs during incubation

Turning or altering the position of the egg during incubation has a definite influence on embryo mortality rate. It is necessary to ensure that the embryo is gently but frequently moved within the egg to prevent it settling and adhering to other structures. In the flat type of incubator it also ensures the more even heating of the egg contents, a factor that does not arise in the cabinet type machine. The need for frequent turning appear to be greatest in the earlier stages of incubation, before the full development of the extra-embryonic sacs and their fluids, but it is the less essential throughout the first 38 days in the setter. The eggs in the hatcher compartment during last 3 to 4 days of incubation do not require turn. The interval between turnings can be as short as 15 minutes, provided that the eggs are turned in opposite directions each time; if the turning is always in the same direction it seems to interfere with the centring action of the chalazae, and embryonic mortality is high. In actual practice it is not necessary to turn as often as this, but it is necessary to arrange the turning with uniform intervals.

Turning in commercial force draft incubators are entirely automatically controlled and the interval may be from 15 minutes to 4 hours. If turning is done by hand, the number of turnings may be reduced to three or five per day; it is important always to turn an odd number of times under such circumstances to ensure that the embryo does not spend the long unturned period always on the same side. Eggs in still air incubator are usually turned by hand, to avoid confusion "o" and "x" marks should be made on either side of the egg in order to confirm the turning effectively. Though there are a few makes of these incubators that have incorporated mechanical device for turning, and the eggs are rotated in long axis. In those cabinet machines that involve traying flat, the same principle is employed, but in most of the cabinet machines where the eggs are trayed in the vertical position, turning is affected by through 40 to 45 degrees in each side. Attention has been drawn to the need to avoid shocks and jarring to hatching eggs during collection and storage. It is even more important to avoid any such disturbance of the eggs in the first 24 hours of incubation, when the proper development of vitelline blood vessels from the blood islands can be inhibited. This would result in about 60 per cent of embryos dying in the second and third day.

Egg weight loss

It is estimated that the weight loss of the egg from fresh weight (immediately after collection) to pip weight (when the air sac has been entered) is approximately 15-18 percent, when the embryo is growing properly. This process can be roughly monitored by weighing the egg. Hatching eggs will provide you with the information needed to set the optimum humidity level. Weighing the egg and keeping accurate records is highly recommended for several reasons. If trouble shooting problems are noted in hatchability, these records will provide you with a good reference on the performance of the incubator. In addition, the eggs of some hens do not lose weight as readily as others, while others lose weight too quickly. This information can be used to determine what action may be taken to compensate, such as increasing or decreasing humidity level or extending or decreasing length of hatching egg storage. The weight loss of an egg is not constant during the incubation cycle. A higher weight loss is noted at the beginning and end of the incubation cycle. This is due to the fact that early in incubation more water must be lost from the albumen in order to provide space for oxygen diffusion to the blastoderm. Late in incubation, water production (and loss) must increase as metabolic rate increases proportional to the larger embryo size. A low relative humidity must be maintained to allow this vapor loss to occur.

The temperature of the incubator should in all cases be kept as stable as possible. Fluctuating temperatures can be detrimental to the hatchability. Proper temperature controls have been designed to keep these variations to an absolute minimum. Daily readings of incubator should be made to ensure accuracy and stability. The dry bulb temperature is directly proportional to the rate of embryonic development. The dry bulb temperature also proportionally determines metabolic water production, oxygen demand, and water vapor pressure within the egg.

The humidity also contributes to the control of the rate of gas exchange of an egg during incubation. The higher the humidity, the lower the rate of gas exchange because high humidity slows the release of water from the egg. Gas exchange is the process of oxygen entering the egg and carbon dioxide and water vapour leaving the egg. This is how the embryo breathes. Eggs should be weighed every 5 days to determine if adjustment of the dry and wet bulb temperatures is necessary for a normal hatch. For example, if a freshly laid egg weighs 1500 grams, calculate the ideal weight loss as 15 percent (15% x 1500g = 225g total target weight loss). If the eggs are incubating at 97.5 °F, estimate that the egg should hatch in about 42 days, then the target daily weight loss = 225 g in 42 days i.e. 5.35g/day. However, it should be remembered that weight loss is not linear because more weight is lost at the beginning than at the end of incubation.

Ventilation

It has already been seen that, apart from the very early stages of development, the embryo is dependent for its supply of oxygen on the air that surrounds it in the nest or the incubator. It is important that the carbon-di-oxide and other gases produced by metabolism are removed from the vicinity of the egg so, that they never reach a toxin level and damage the balance of gas exchange. Ventilation of the incubator has then, the dual object of providing the necessary oxygen and of removing excess carbon-di-oxide at a sufficient rate. An ordinary chicken egg requires about 5 liters of oxygen to enable the embryo to mature, and in the same time it gives off about 3 liters of carbon-di-oxide. The oxygen content of fresh air is around 21 percent, which is also the optimum concentration for the development of the hatching egg. It has been shown that the oxygen level can be reduced to17.5 percent without affecting hatchability but, reduction to 15 per cent reduces embryo survival. As long as there is continual intake of fresh air and expulsion of used air there will be no risk of oxygen starvation under normal atmospheric conditions. For this reason, it is vital to make sure that the incubator room is thoroughly well ventilated, since it is the source from which the incubator draws its air.

Hatching

It is recommended that eggs should be transferred when a chick pips into the air cell internally. This can be confirmed by candling; further the hatch time can be more accurately predicted by candling. In high humid area, the relative humidity is difficult to control, in that case, the embryo should be allowed to pip externally before moving to the hatcher. It is suggested that the hatcher be maintained at the same relative humidity as that of setter until at least external pipping has occurred.

This will allow the egg to lose the water necessary to get the proper oxygen intake during this very critical period of embryo development. During the final stages of the hatching process and after external pipping, a relative humidity of 30 to 40 percent is recommended. It is also suggested that the hatcher dry bulb temperature be run at 0.50F higher than the setter temperature since the hatcher will routinely have fewer eggs than the setter.

(a) Normal Hatching

Ideally, 85% of the eggs should hatch on their own, or with will minor assistance. Candling of eggs should be done on day 39 of incubation. When the drop of the air cell membrane begins, this is the first sign the chick is beginning to position itself to hatch. At this stage, which will last approximately 24 to 48 hours, candling every 12 hours is adequate. If the desired weight loss has been achieved and the eggs have been incubated properly, unassisted internal pipping is the rule.

As the chick positions itself we will see progressive movement at the air cell membrane, which exhibits itself as uneven protrusions of the usually smooth membrane. This movement is the beak and feet pushing against the membrane. By gaining experience a farmer will recognize correct, as opposed to incorrect, movement. When internal pipping is confirmed through candling, we will observe a darkening of the air cell due to the chick entering the air cell. At this time it is important to note the time of internal pip. As a rule of thumb 24 hours is the optimum time from internal to external pipping. However, if movement continues to be vigorous without a significant change, the 24-hour window can be lengthened with careful observation. As the chick rotates, it uses a muscle in its neck (appropriately termed the pipping muscle) to break the egg shell. The matrix of the cellular structure of the shell makes it easier to break from the inside than from the outside. From the external pipping point to the completion of the hatching process, approximately 12 to 24 more hours should elapse.

After hatching, the chicks should be allowed to remain in the hatcher only for the time sufficient to provide for drying without dehydration. A chick that is up and moving about is ready to be removed from the machine. On the average, the holding time after hatch should be about 12 hours.

(b) Assisted Hatching

Time and experience will teach whether a chick needs assistance or when to let nature take its course. It is best advised not to allow a chick to remain in the air cell for more than 24 hours without intervention. Stress can cause many problems in a neonatal chick. The fact that a chick has internally pipped does not necessarily mean that it is in a position to continue the hatching process. It must be stressed at this point that if more than 10% to 15% of eggs need assistance at hatching time, it may indicate a problem in the incubation process or with the reproductive health of the hen.

How to assist hatching?

The egg may be placed on a towel, on a well-lighted level surface. The air cell end of the egg should be towards the handler, assuming the chick's beak should be located approximately at the 10'o clock position.

- Using a blunt instrument such as a tack hammer, or other similar object, the egg can be tapped until a small crack appears.
- Using a pair of tweezers to break very small pieces of shell away, until the broken area has a hole about the size of a quarter in the egg.
- The chick is known to be in the air cell at this point. At this stage the bird chirp can be heard. At this point the bird's beak can be seen.

In some cases a chick may appear to be in the air cell when in fact the inner membrane has collapsed around the beak.

If this is the case,

- Then the shell may be broken further to be able to reach into the egg and the membrane is torn.
- Using a clean tweezers or small, sharp scissors (gently without damaging the blood vessel) the membrane may be pulled away just enough to expose the beak.

Once assured that the nasal passage is clear and the chick is breathing, the hatcher can be closed and the chick is left alone for at least two hours. Slow and cautions are the key words here. If after two to three hours the chick has made no progress, shell may be removed further to extract the head and neck completely from the shell so, that no aspiration occurs.

Unhatched chick after internal pipping

(Courtesy: PGRIAS, Kattupakkam, TANUVAS, Tamilnadu, India)

Trouble shooting chart during Incubation and Hatching of Ostrich

Condition	Probable Cause	Remedial measures
Fully formed chick, dead without pipping	1. Improper turning 2. Malposition 3. Inbreeding 4. Incorrect temperature	1. Increase turning 2. Tilt sac end up 3. Use unrelated pair 4. Use correct temperature
Chick dead in shell after pipping	1. Low average humidity 2. Low average temperature 3. High temperature surge	1. Increase humidity 2. Raise temperature 3. Protect from surge
Mushy chick, dead after hatch	1. Infection in hatcher	1. Disinfect hatcher
Sticky chicks	1. Low average temperature	1. Increase temperature
Shell sticking to chick	1. Low average humidity	1. Increase humidity
Chick smeared with egg contents	1. High average humidity 2. Lack of Oxygen	1. Lower humidity Increase ventilation
Rough navel	1. High temperature 2. Wide variation in temperature 3. Low humidity	1. Lower temperature 2. Hold temperatures teady 3. Increase humidity
Bloody navel	1. Temperature too high	1. Lower temperature
Large soft bodied chicks	1. Low average temperature 2. Poor ventilation	1. Increase temperature 2. Increase ventilation
Hatching early	1. Temperature too high	1. Lower temperature
Hatching late	1. Temperature too low	1. Raise temperature
Short down on chicks	2. High temperature	1. Lower temperature
Chicks dead at 5 - 6 weeks	1. Poor ration 2. Lack of ventilation	1. Correct ration 2. Increase air flow

Dead in shell chick due to unabsorbed yolk

(Courtesy: PGRIAS, Kattupakkam, TANUVAS, Tamil Nadu, India)

❑❑❑

Chapter 14

NUTRITION AND FEEDING MANAGEMENT

Ostrich birds are monogastric animal and their nutritional requirements are formulated based on this aspect. The most important nutrients are energy, protein, minerals and vitamins. These should be provided in correct and accurate form in different stages of growth to expect the desired effect. Since most of the growers are rearing the birds in large sized paddocks/ ranches, apart from the grazing pastures, they need to be supplemented with concentrate mixture.

Digestive Physiology

The digestive physiology of the Ostrich is unique amongst avian species. Compared to the domestic fowl, the gastrointestinal tract of the Ostrich is without a crop, has extremely well developed sacculated caeca and unlike other ratite birds has an unusually long colon. The tongue of the Ostrich is blunt and folded back over itself rostrally, forming a pouch or pocket. The oesophagus generally traverses the right side of the neck, but all cervical structures are movable. It is of great importance to note that the crop is absent in all ratites. The oesophagus enters the stomach (proventriculus) within the thoracic cavity.

The distal end of the proventriculus passes dorsal to the ventriculus (gizzard). The opening between the proventriculus and ventriculus is large enough to make it possible to extract any foreign bodies in the ventriculus via an incision in the proventriculus. The ventriculus of the Ostrich is a thick-walled structure similar to the ventriculus of seed-eating birds. The dark tough lining (koilin) of the proventriculus and ventriculus is formed by protein secreted from the glands combined with entrapped sloughed cells and cellular debris.

A duodenal loop is formed, with the pancreas lying between the segments. The small intestine is relatively short and straight. The rectum in the Ostrich is voluminous and occupies the caudal right abdomen. Such a long and large rectum is necessary for digestion of bulky food items and fluid absorption. The coprodeum and urodeum are partially separated by a copro-urodeal fold. The urodeum is short but within it are orifices for the ureters, the oviduct of the female and the vas deferens of the male. An uro-proctodeal fold partially separates the urodeum from the proctodeum through which all excretions pass via the vent to the exterior. The bursa of fabricius is on the dorsal of the proctodeum.

Nutrient requirements

Ostriches are able to make better use of energy locked up in forages than poultry. Digestibility of Apparent Metabolizable Energy (AME), fats and Neutral Detergent Fibre (NDF) increases with the age. Since, Ostriches are not typical poultry but monogastric herbivore, the ME values are 41% higher than that of poultry.

Feed digestion

The digestive system of the Ostrich is that of simple stomached herbivore, which has the ability to utilize the forage, adult Ostrich has a hind gut fermentation like the donkey, horse, turkey or rabbit. Therefore, it is essential that their intestine are inhabited with useful bacteria as early as possible. The gastro-intestinal tract (GIT) starts from the mouth in which the teeth are replaced by a beak. The mouth empties to the upper part of the oesophagus which is pouch like that allows the accumulation of feed. The feed moves visibly down the gullet when swallowed and empties directly into the glandular stomach (proventriculus), where digestive enzymes leads into the gizzard which usually contains grits. These grits are used to grind the feed into smaller particles, where the digestive enzymes play their role of digestion. The grits are usually worn down, gradually and are never exerted whole. The stones offered to Ostrich chicks should be round in shape and not angular stones to prevent damage of the GIT by the sharp ends. These stones should be made of strong material such as quartz and not substances that easily dissolve it like oyster shell, limestone or dolomite.

The gizzard empties the ingesta into the small intestine which is relatively short and leads into the colon. Two large caeca protrude from the end of the small intestine. The colon is very long and may be longer than 10 meters in adult bird. Microbial fermentation takes place in the hind gut which leads to the formation of volatile fatty acids (VFA) which, are absorbed and metabolized as an energy source. The Ostrich is capable of digesting fibrous feedstuff such as cellulose and hemicelluloses. The excreta pass from 21 to 76 hours after consumption of feed. Further, the hindgut provide best microbial environment for microbial fermentation micro flora.

The natural diet of Ostriches is mainly green grasses, berries, seeds, succulent plants and small insects. The adult Ostrich on pasture eats three times as much dry matter in relation to the dairy cow when expressed as a percentage of body weight (7.5% vs 2.5%). When practicing a system of alternative grazing, the carrying capacity of lucern pastures is 8 Ostriches per hectare. Cost of feeding is the largest expenditure in Ostrich farming. A well conditioned chick with optimum growth rate and well nourished is less likely to succumb to diseases. Nutrition plays vital role in Ostrich production. Ostriches have very efficient feed conversions during the first 210 days of their life. After 330 days they become inefficient and their feed conversion drops. Therefore, it is important that chicks are fed optimally when young so that they attain the ideal slaughter weight of 90-100 kgas quickly aspossible.

Green fodder supplementation for the adult birds

(Courtesy: PGRIAS, Kattupakkam, TANUVAS, Tamil Nadu, India)

Chicks learn to eat under natural wild condition by copying their parents feeding behaviour. Therefore, under commercial artificial production system, a surrogate mother or an older chick should be used to teach the young chicks how to eat. Chicks depend on their yolk sac for the bulk of their nutrition in the first 10 to 14 days of life. They should therefore be eating well by the time the yolk sac is finished so that the chicks maintain normal growth and development. Grit and feed should be offered to the young chicks. Concentrate offered to the chicks should contain 20% by weight of green forage as the green colour stimulates pecking.

Shallow troughs or trays should be used to offer chicks their feed. Palatability of the feed is enhanced when it is dampened. Ostrich feeds can be in pellet or mash form. When feed is offered as mash, it reduces cost since pelleting of feed incurs higher cost. Birds spend more time pecking at the meals and therefore, develop fewer behavioural disorders. Farm mixed feeds cannot readily be pelleted hence, molasses or stabilized fat can be added to reduce dustiness and increase palatability. Pellets are easier to be used as feed. This results in less wastage, minimal dustiness and less separation of feed ingredients. This is however more expensive than the use of meals. Commercial farm prefer pellet feed over mash feed due to the above advantages.

Proper feeding and nutrition is not only a vital part of a successful management programme but, also it is necessary that Ostrich birds need balanced diet containing full complement of ingredients which will be able to provide required level of protein, energy, vitamins and mineral levels for proper growth and reproduction. Since feed cost alone contributes 60-70% of the production cost, the least cost rations will improve the margin of returns. Feed can be prepared by using common poultry feed ingredients as per the suggested nutrient requirements at different ages and during breeding and non-breeding stages.

Ostrich chicks usually do not feed anything during the first 7 days. The yolk sac will provide sufficient nutrition during these days. High fiber food will initially cause impaction and result in death of chicks. Only clean and fresh feed should be fed to the chicks. Among the chicks "starve outs" are quite common. Continuous feed and light should be accessible to the chicks. Greens and other high fiber diets can be fed gradually and in small quantities. Clean water is also equally important for good chick growth. Chicks can be grown by feeding good quality feed that has been fed on continuous basis. The chicks may be initially fed with a diet containing 22 % protein and later from 3 months onwards until 18 months fed with 16 to 18% protein. Green fodder should always be made available. There after maintenance diet of 16% protein may be fed until sexual maturity. During laying a protein of 18 to 20% will be sufficient.

Major portion of the birds feed consists of carbohydrates which serves as energy source and proteins which has important function for the building body tissues during growth and reproduction. Both water and fat soluble vitamins are required in small quantities for proper reproduction, increased disease resistance, maintain integrity of the epithelium, assimilation of calcium and phosphorus from the feed, preventing rickets and leathery or thin shelled eggs and also for the normal growth. Minerals are needed for deposition in skeleton, in egg shells, and even for feather pigmentation and to prevent skeletal abnormalities and certain other metabolic functions.

Like chicken, Ostrich also eat small stone pieces for digesting food properly. Adult Ostriches can survive without water for a few days. They collect water from their consumed grass or leaves. But if they find water, they drink and often take a bath. Ostrich can tolerate frequent changes of temperature. Even they can tolerate daily temperature change of 40 °C. During excessive cold weather, they cover their legs with their large wings, which helps to keep their body warm.

Water Requirement:

Clean good quality water should be made available at all times.The importance of water is felt more in Ostrich rearing than in chicken. Water usage is approximately 2.5 lit per kg of dry feed intake. The importance of water is felt more when the birds are reared intensively. Since, the chicks are mostly reared under intensive conditions; importance of quality water should be given top priority. Types of weather, room temperature, environmental conditions, type of feed, quality of water, production stage, housing, etc. play a role in water requirement. Birds reared under extensive or free range system, seldom need to drink water as the same is met by their grazing of succulent plants.

Common components of Ostrich feed

Concentrates	Protein sources	Mineral sources	Roughages	Others
Maize Wheat Oats Barley Distillers grain, etc	Soyabean cake Sunflower cake Canola cake Fish meal Meat meal Full fat soya Sunflower seeds Beans Peas, etc	Di–calcium phosphate Mono calcium Phosphate Mineral mixture Lime stone	Lucerne Wheat Oat Silage Wheat bran Rice bran, etc	Syn. Lysine Syn. Methonine Molasses Growth promoting agents Pre and Pro-biotics

Adult birds being supplemented with powdered green fodder over the feed
(Courtesy: PGRIAS, Kattupakkam, TANUVAS, Tamil Nadu, India)

Chick Feeding

It is recommended that Ostrich can be started with a diet containing minimum 22% crude protein. This has been found to give good results in rearing the chicks. Some practice by withholding the feed up to 72 hours in the newly hatched chicks. This allows complete absorption of the yolk sac. This is debatable and is left with the individuals. Either can be practiced. The suggested commercial guidelines for nutrient requirements for Ostrich at different stages are furnished in the below mentioned table. The starter feed should be formulated to contain all the nutrient requirements to support optimum growth rate and health status of the Ostrich chicks during the first two weeks of their life and feed can be offered in crumble form and all other ages can be provided with pellet diet. Young birds lack gastroliths and microbes necessary for digestion of feed. Hence, they tend to swallow small stones to aid in digestion.

It must be ensured that chicks will have continuous access to the starter feed at all times of the day during the first three weeks of period. In order to ensure proper feed intake and to support the above continuous availability of light should be ensured. Feeding of higher levels of crude protein will lead to overweight which may lead to problems such as leg weakness, leg abnormalities and death.

Even though the fiber requirements are more than the other poultry species, higher dietary fiber level is reported to cause "Starve out deaths" due to intestinal obstructions in young chicks. Always take care to see that chicks are never fed mould or musty feeds or high moisture content feed as this may cause mycotoxicosis, botulism, mortality problems and decreased production performance.

Management during early rearing period will definitely has a bearing on the growth and health status and production performance of the adults. Maximum care should be exercised in management and feeding of chicks. Nutrition related problems in chicks and juveniles include the following: leg weakness, head retraction, bowed or bent leg bones, splayed legs, intestinal obstruction etc., One should always remember that no single ingredient is sufficient to prepare an ideal balanced diet for feeding Ostrich to meet the nutritional requirements.

Minimum nutrient requirements for commercial feeding of Ostriches (Open feed formula)

Nutrient	Pre starter (0 – 2 m)	Starter (3 – 6 m)	Grower (7-12m)	Finisher (13-14m)	Maintenance (>15m till sexual maturity)	Breeder (Laying)
Min. Crude protein (g/kg)	190	170	150	120	100	120
Min.Crude fat (g/kg)	25	25	25	25	20	25
Max. Met. energy (Kcal/kg)	2700	2800	2800	2600	2600	2500
Min. Lysine (g/kg)	10	9	7.5	5.5	3	5.8
Max. Moisture (g/kg)	120	120	120	120	120	120
Max. Crude Fiber (g/kg)	50	50	70	80	100	130
Min. calcium (g/kg)	12	12	10	9	8	25
Total phosphorus (g/kg)	6	6	5	5	5	5

Grower feeding

At 12th weeks of age, the Ostrich chicks should be fed with grower feed and their feed should be split and fed in two halves. Juveniles should be switched over to maintenance diet at 25 weeks of age and continue with the same diet until they attain sexual maturity. Breeder ration should not be fed to the juveniles as it contains higher levels of calcium. These birds should be started with a breeder diet two weeks prior to the onset of laying egg.

While changing from one type of feed to another, it must be introduced gradually. Initially 25 percent of the new diet is mixed with 75 percent of the already used diet. After four days, equal quantities of old and new diets can be mixed. After 8 days 75 percent of the new diet can be mixed with 25 percent of the older diet. After two weeks, only new feed can be given. This process is done for the birds to take a slow transition, because, if changes are made suddenly, birds may avoid the feed or they may develop diarrhoea.

Adult feeding

Adults and older juveniles can be ranged on well drained, succulent pasture plots or they can be maintained under dry-lot conditions. Pastured birds should be maintained on small plots and rotated periodically for optimum grazing efficiency. Birds on large acreage may become semi-wild and difficult to manage. Succulent forages such as irrigated alfalfa, wheat, rye, oats, or bermuda grass are desirable for juveniles and adults and it will reduce production costs significantly. Some forages may be too tough or high in fibre or too low in protein and energy to provide the nutrient level required for desired growth performance. If Ostriches are allowed to forage on natural grasses or grain, balancing the nutrient intake with the sufficient amount of commercial feed is advisable. Ostriches should not be allowed to become overweight. Excess body weight in the form of fat is detrimental to egg production and for slaughter (meat quality). Breeding birds on pasture should receive daily supplements of a good quality, high protein breeder ration to ensure optimum egg production, fertility, and hatchability. Breeders in dry-lot confinement also should be fed the ratite breeder ration with continuous supplemental feeding of good quality alfalfa pellets or cubes of hay. Pellets or cubes are less wasteful and easier to feed than hay. Breeders should have access to oyster shell, grit, or commercially prepared bone meal during the breeding period. The addition of poultry vitamins and electrolytes to the feed or water may improve fertility and hatchability.

Feeding behaviour of Ostriches:

Ostriches spend 70 to 80% of the day feeding with continuous ingestion and movement being important in their digestive physiology. In the wild, the diet consists of 60% plant material, 15% fruits and legumes, 4-5% insect eggs and small mammals, the rest consisting of cereal grain, salts and stones. Feed should be made continuously available to the Ostriches rather than fed as meals. The young Ostriches displayed an innate preference and showed little change in diet composition with increasing experience. The most heavily utilized foods are newly emergent grass leaves and the foliage of the woody shrubs. Ostriches are active for about 12 hours during the day. Research conducted at this unit suggested that, Ostrich spent about 10% of their time feeding, 20% pecking at the soil, 50% resting and around 2% drinking water. The feed conversion ratio for the Ostriches is higher than that of the other poultry species. Ostriches feed only during day time and are mostly inactive during night time.

The following forms of feed can be recommended for the different stages of growth

Age / Stage of the birds	Type of recommended feed
Pre – starter feed	meal / Crumless
Starter feed	Crumbles
Grower feed	Pellets (Maximum size 5 mm)
Finisher feed	Pellets (Maximum 8 mm)
Maintenance feed	
Breeder diet	

Importance of feed cost reduction:

Since, the feed cost involves 70 to 80% of total cost, more attention should be given to the feeding management. As these birds are highly sensitive to nutrient requirements, maximum attention should be given to feeding. The following maybe followed scrupulously;

- Minimizing feed wastage.
- Proper feeding management
- Preparing own, good quality cheaper feed

Actual requirement of feed alone should be provided and never more than that should be offered. Force feeding is not possible, but it should be avoided at any cost. Increased feed intake would result in increased cost of production per unit. Besides this will also result in the birds continuously picking up certain feed components selectively and in this process certain constituents are excluded from the feed which, may lead to deficiencies in certain birds. Feeding of Ostrich with pellet feed is recommended and it has several advantages. Feed offered should be accurately weighed and their records should be maintained to draw the future feeding programme based on the defects observed from the earlier feeding records. Forages such as Lucerne, Stylo, Desmanthus and other perennial legumes are preferred for juveniles and adults which will also help to reduce the cost of production and improve the economics. Locally available fodder can also be selected, if preferred by the Ostriches. Some forage may be too hard or high in fiber content or too low in protein and energy to support nutrient level required for desired growth. It is important that quality forages are selected to provide better nutritional status to the Ostrich birds. When the Ostriches are allowed for free foraging on natural grasses or grains, they will try to balance the nutrient intake with the amount of commercial feed that is fed.

The feed formula cannot be fixed and cannot remain the same for different parts of world. It depends upon the type and quality of individual ingredients, method of growth, types of pasture, soil composition, type of water used for irrigation, etc. Maximum growth in the birds is attained when they are about six months of age. There after the feed consumption increases but not corresponding to the body weight.

The following would indicate the expected feed intake and growth rate of different stages of Ostriches

Stage of growth	Age (months)	Average daily weight gain (ADG) g/bird/day	Feed intake (g/bird/day)
Pre - starter	0 – 2	210 – 240	680 – 720
Starter	3 – 6	410 – 440	900 – 1100
Grower	7 – 12	390 – 400	1500 – 1700
Finisher	13 – 14	220 – 260	1900 - 2200

Feed intake, body weight and feed conversion

Wide variations have been reported with regard to the feed consumption; body weight gained and feed conversion efficiency around the world. The observations made on the Ostrich breeding unit maintained at the PGRIAS, TANUVAS, Kattupakkam, Tamil Nadu, India are given below. The production particulars of Ostriches are subjected to variation depending upon the purpose for which they are maintained.

Feed consumed, body weight gained and feed conversion ratio

Age in months	Feed consumed	Body weight (kg)	Feed Conversion Ratio
0	--	0.850	--
2	18.0 – 19.0	12.0 – 13.0	2.15
4	28.0 – 29.0	31.0 – 32.0	2.46
6	53.0 – 54.0	51.0 – 52.0	3.32
8	67.0 – 68.0	71.0 – 72.0	4.23
10	73.0 – 74.0	88.0 – 89.0	5.05
12	78.0 – 79.0	104.0 – 105.0	5.77
14	81.0 – 82.0	109.0 – 110.0	6.99

Chapter

15

HEALTH MANAGEMENT

As long as Ostriches were reared under open system, the health of the birds was not a major concern. Now, the Ostrich rearing has increased and they are being reared under closed system. The need of the hour is to rear the birds profitably, which has also resulted in the health of the birds. The health of the bird has now great influence on the profitability of the farm. Hence, it is mandatory to follow a suitable health program which when followed would result in better and higher profitability. The health program is also influenced by the types of rearing one is likely to follow. Some farmers rear the birds as growers only and some rear as breeders and some combine all the facets of production including hatching of chicks. Unlike in poultry, the health program can be separated into three main categories via, the bird itself, the type of management including the environment and finally the agent which causes the diseases. It is not enough, if one part alone is subjected to the treatment. While taking care of the main causative factors, other two should also be probed and rectified. Some of the characteristics features of a healthy bird are given below as a general guideline;

- Symmetrical gait with wings and legs
- Always curious and vigilant looking
- Head and neck is erect and held high
- Spends maximum time in pecking and browsing in the ground
- Always found within the group
- Well groomed feathers
- Voids dark pelleted dung first and urine later
- Never sits down, always in standing position
- Have a bouncy walk like camel.
- Good appetite for food and water

Any deviation from the above is construed to be sick birds and need special attention.

Problems encountered during brooding phase

The foremost procedure performed upon the chick is care of the navel, because the blood vessels which connect the structures of the egg membrane to the internal structures of the chick can act as a wick along which viruses, bacteria or fungi are likely to travel. Hence, taking care of navel is important;

1. The area around the navel should be trimmed with scissors and sprayed with 7% iodine.
2. A gentle traction on the dried wick is placed and it is cut with sterile scissors at skin level.
3. The area is sprayed with 7% iodine for a second time.
4. A small dab of Betadine ointment can be placed at this point.
5. A small (2″ x 2″) gauze pad is placed on the navel.
6. Using a vet wrap or masking tape around the area will hold the dab in its position.

The chicks usually stay in the brooder for eight to twelve hours, until they are dry and they have become active.

Blood collection in chicks

(Courtesy: PGRIAS, Kattupakkam, TANUVAS, Tamil Nadu, India)

Importance of temperature

Temperature is extremely important for all domestic birds, but is particularly so for day-old chicks, which are exposed to temperature stresses even before they are placed in the brooder/grower house. For instance, temperature variations in the hatchery holding room and during transport will often severely stress the young chicks. Careful control of the temperature is an absolute must during this period of transition. Much has been made of possible chilling of baby chicks. If one but thinks back, it is obvious that the temperature inside the egg was 96.5°F for about 25 days and then gradually increased to about 99.5°F. When the organs of the embryo begin to function and the metabolism of the embryo occurs, the temperature within the egg is approximately 3 degrees hotter than the temperature setting of the incubator. The temperature then drops from 99.5 to 96.5°F when the egg shell is cracked and to 95oF in our brooder. We keep the outside brooder at 85 to 90°F for the first 36-48 hours, because the chick's ability to regulate body temperature is limited for the first few days. Apparently a transition to temperatures of 75-95oF is tolerated very well after 48 hours.

The chicks are usually placed down on the ground by the second day because early activity is extremely important and don't wish them to become lazy and docile. Chicks are not given feed or water until they are placed down. The temperature at which the Ostrich chick emerges from the egg is between 35 and 36 °C. It is logical therefore that the brooding temperature, during the first two days, should be as close to 35 °C as possible to avoid too much of a change in the chick's first environmental experience. Chicks should be started at 33 to 34 °C and the temperature should be lowered gradually (either daily or weekly) until an ambient temperature of 21-23 °C is reached at about four weeks of age.

The second period of the chick's life is that time from the first outside exposure until it is placed down on the substrate and possibly mixed with older chicks, this period is usually between 12 to 18 hours. Once the chicks are dry, their bandages in place, their microchips implanted and observation indicates that their general health is good, they go directly from the hatcher-brooder to a larger brooder in the chick barn.

The approximate dimension of the brooder are 2′ x 2′ x 8′ and the floor is covered by tightly-woven indoor / outdoor carpet. It is heated by thermostatically controlled coils and by movable hat lamps.

(a) Yolk-sac retention

Weighing chicks daily in the morning is a tool that can be useful in making decisions on yolk sac retention and future course of medication. Ostrich chicks should gain 10 to 15 g each day after 5 – 7 days. Fluids given at this time can help prevent dehydration, which can lead to lethargy and ultimately, death. Caloric intake can provide energy. High-fat content foods are not recommended for this situation, since fat is not a digestible substance in a starvation situation. The biggest threat to the Ostrich chick at this time of life is to get into a weight deficit situation. They seldom recover. If a chick does not respond to recommended treatment within 24 hours, yolk-sac retention should be considered. The chicks under normal course should lose weight till 5 to 7 days and thereafter only gain weight.

(b) Omphalitis

Most of the mortality that affects the chicks is either of intrinsic aetiology pertaining to the un-absorbed yolk sac or omphalitis or generalized oedema of the post hatch. Other mortalities are of extrinsic aetiology like heat stress or gut impactions due to the non-selective nature of the starter chicks. Improper management lead to losses due to drowning or severe injuries.

(c) Spread legs

Improper incubation leads to this problem and most likely it occurs within one weeks of its age. Other actors like floor, ambient temperatures, brooding surrounding, etc should be taken care of.

Chick with leg deformity

(Courtesy: PGRIAS, Kattupakkam, TANUVAS, Tamil Nadu, India)

Problems encountered during growing phase

Grower chick mortalities are all of extrinsic aetiologies related to the environment or to the management aspects. Improper fence constructions can lead to injuries and strangulations. Airborne polythene bags, paper or miscellaneous items may be engulfed by chicks, which may cause impactions or death. Occasionally slippery floor after rains or faulty water run offs can cause leg problems as the weight of the chicks increases. Affections of environment origin seem to be of equal probabilities. The percentages of mortality during chick and grower phase have shown vast variations worldwide. The range may vary anywhere from 10 to 90%. It is highly unassuming to ignore this fact. The causes of mortality are also vague in nature in most places. However, a good management protocol can definitely prevent this tragedy and make profitable enterprises. The commonly encountered issues as detailed below;

(a) Parasites

The prevalence of internal parasites in intensive or closed type of rearing is low compared to open and pasture raised birds. The advantage of artificial hatching is that the chicks are separated from their parents and reared separately. It helps in the breaking up of their life cycle. So, low infestation expected. The ecto parasites like ticks and feather lice are commonly observed in the birds which are reared under intensive system. Most of the parasites are of little economic importance, except the few like Ostrich wireworms (stomach worm), Ostrich tapeworm (Houttynia struthionis), Nematodes (Codiostomum struthionis), Coccidiosis and Eye fluke. These worms can be controlled with broad spectrum anthelmintic medications

External parasites like lice, mites, ticks are most common and widespread. They affect almost all the age groups of birds. The birds will exhibit irritation, scratch and in extreme cases lose of feathers. Lice of genius Struthiolepeurus can cause severe feather damage. The most area of attack of them is head and neck. High infestation can be expected in places where the rainfall is high and dense vegetation is available for Ostrich rearing. These ectoparasites can generally be minimized by sanitary preventive measures.

(b) Physical problems

Impactions

Impactions is one the most frequently observed problems in growing chicks to adult hood. Two types of impactions are recognized namely; acute or newly acquired impactions and chronic or long standing impactions.

Acute impactions usually result from a bird eating too much unusable material very rapidly. This type of impactions can result from intake of sand, dirt, gravel, hard grasses and so on. Death in these cases is often quick due to the over-extended and non-functional condition of the proventriculus. Chronic impactions usually result from the bird having a partially impacted proventriculus which allows some material to pass through normally. These birds do not gain weight normally. They often appear malnourished and tend to eat excess non-food materials. Impaction in Ostriches may result from many reasons such as nature of the feed, stress factors and the environmental factors. Excessive ingestion of hard seeds, dry grasses and wild fruits may lead to impaction in older Ostriches. Consumption of excessive quantity of high quality ration by young chicks, rapid or sudden change in diet, from low to high-fibre rations, poor quality fibre, etc may result in impaction. The animal factors that can result in impaction in the Ostrich are

(i) Chicks left hungry for long periods without feed hence eating a lot of foreign materials, such as sticks, strings, plastic tags, bones or wire.,

(ii) Chicks raised on sandy floors may eat excess sand if the floor is soft or wet,

(iii) Stress leading to displacement feeding and

(iv) Also specific nutritional deficiencies leading to deprived appetite.

Lack of gird and small stones to assist in grinding ingesta in the gizzard may initiate impaction in the birds. Impaction can be prevented by avoiding the major causative factors in a farm. Impaction is not likely to be major problem under intensive bird feedlots but it can be a significant cause of mortality in young chicks under semi-intensive systems where chicks have access to pastures freely or on restricted basis. Chicks previously confined to concrete and turned out to pastures for the first time usually get excited and tend to eat almost anything which may initiate impaction. Early treatment with some form of oral lubricant such as vegetable oil can help to breakdown some of this material and get the bird back to a normal diet. When diagnosed early, surgical interventions would be very successful in removing the impacted material. Oral dosing of Epsom salt (magnesium sulphate) in water teaspoons for small chicks to 2 tablespoons for an adult bird can also help in the treatment of the birds.

Coprophagy

Under natural conditions, newly hatched Ostrich chicks consume their parent's fresh faecal material which may be the source of micro-organisms that reside in their digestive tract.

Feather pecking and cannibalism

Feather pecking in birds can be corrected by feeding Ostrich lick at 100-150 g/ bird/day. Ostrich lick contains 25% protein, 0.3% Methionine, 1.36% lysine, 1.67% Arginine, 0.25% Tryptophan, 7.0% Fibre, 6.97% ME, 3.9% Calcium, 2.1% Phosphorus and 5% Salt (Johnson, 1999).

Obesity

Excess energy above the bird's requirements for maintenance, growth or reproduction leads to obesity. The thickness of fat in the ventral abdominal wall is a good indication and may vary from nil to 8 cm in thickness. Condition scoring of the birds can be carried out (scale 1 to 4). This is done on the curvature of the back viewed from the side, an amount of fat bulging around the tail and the shape of the abdomen viewed from behind. The spine over the thorax, can be palpated just anterior to the sacrum. Obesity may lead to decreased libido, low egg production and susceptibility to heat. Obesity can be controlled by monitoring feeding of the birds.

Leg rotation

Leg rotation is a major cause of chick mortality with an incidence of 50 percent the leg injuries. Contributing factors are thought to be genetic susceptibility, poor incubation control, overfeeding, slippery floors and inadequate exercise. Trauma to the neck and limbs also occurs frequently.

Trauma

Trauma is the single biggest cause of death in juveniles from three months of age. Fractures, lacerations, dislocations of limbs, wings are commonly encountered during improper handling of birds. Incorrect paddock design, incorrect fencing design, materials, improper distribution of birds, etc add up to the existing issues.

Bacterial diseases Omphalitis and yolk sacculitis – discussed earlier

Botulism

It is caused by the consumption of toxins released by Clostridium botulinum. Loses are mostly by type C toxin. Drowsiness, weakness, etc are earliest signs. Administration of type C antitoxin can bring about total recovery in birds.

Anthrax

Ostriches are the only birds susceptible to anthrax. The disease may be per acute or acute. It is caused by *Bacilius anthracis*. It is commonly found in places where the humidity is higher than 70% and temperature is in the range of 35 to 42°C. Transmission is mainly be spores. Pencillin, if given early can be very effective.

Necrotic enteritis

This condition is caused by *Clostridium perfringens*. It mostly affects the young Ostriches especially under 8 weeks. It quickly spreads through the droppings of the bird. Any broad spectrum antibiotics would be effective.

Campylobacteriosis

It is a very severe diseases caused by *Campylobacter jejuni*. They commonly affect the birds of age 2 weeks to 16 weeks. Production of green stained urine is a major symptom observed. Any broad spectrum antibiotic would be effective, if treated early.

Prolapse of penis in adult male Ostrich
(Courtesy: PGRIAS, Kattupakkam, TANUVAS, Tamil Nadu, India)

Viral Diseases

New Castle disease

It is highly contagious and lethal disease affecting all age groups. It causes respiratory and nervous disorders in the birds. In acute form, it can kill all the birds and spreads faster to unvaccinated flocks. Vaccination is the only prophylactic measure for this condition.

Avian Influenza

The well known HPAI (Highly Pathogenic Avian Influenza) is equally highly contagious and lethal disease affecting all age groups. In young birds it can cause more than 90% mortality and about 70% in older birds. No vaccination and no treatment are available currently.

Nutritional disorders

Alteration of nutritional balances can cause mild to severe conditions in birds. Deficiency disorders are commonly found in Ostriches, if they are reared in pastures and have very little access to concentrate feed. But, Ostriches are such a wonderful bird, that once the deficiency is identified and supplemented, the symptoms vanish immediately. Identification of the deficiency of the micro or macro or both nutrient needs to be done first. Though the levels of requirements are not as standardized as in chicken, but still the chicken requirement level can be taken as base value. Essential vitamins required are Vit.A, Vit.D, Vit.E and Vit. B-complex. Optimum mineral supplements are required to avoid deficiency conditions. Minerals like manganese, copper, zinc, etc are vital for regular body functions.

Transmission of infectious agents from other species to Ostriches has to be totally prevented/minimized. The most important sources of infectious agents in Ostriches are domestic and wild birds. Whenever possible Ostrich farming should not be undertaken in the vicinity of a poultry farm, and no poultry or pet birds should be kept in the vicinity. Probably the most important starting point while dealing with a medical issue in Ostriches is to obtain the most accurate and extensive clinical history (anamnesis) of the flock, and the individual, before starting with the clinical inspection of the sick birds. The most common medical problems observed in Ostriches are in some way management or nutrition or both related, which makes it imperative for the Veterinary Practitioner to have good basic knowledge on every aspect related to Ostrich nutrition, management and handling. Factors such as incubation and hatching conditions, as well as over-crowding, environmental temperature, handling and stress, position of drinkers and feeders, kind of litter, exercise space, changes in diet, and many other factors may be related to medical problems within the flock. All possible factors should always be checked while investigating a clinical case in Ostriches.

Mortality pattern observed at PGRIAS, Kattupakkam, TANUVAS, Tamilnadu, India)

Mortality of Ostriches as observed at PGRIAS, TANUVAS, Kattupakkam revealed that about 70% death among the total mortality occurs during the first two months of chick's life and as the age advances, the mortality reduces considerably. As this has been cited by various authors, early chick mortality can go as high as 80%. Hence, more emphasis should be made on the chick management. The details of death in Ostriches as observed are given below:

Age, percentage and causes of death in Ostriches

Age (months)	Percentage of death	Major causes
0 - 2	60 – 70	Omphalitis, impaction, starvation, enteritis and septicemia
2 - 4	10 -15	Impaction, starvation, enteritis, septicemia and leg deformities
4 - 6	10 -15	Impaction, starvation, enteritis, septicemia and leg deformities
6 - 12	< 5	Cannibalism, traumatic injuries, starvation, enteritis and impaction
> 12	< 5	Cannibalism, traumatic injuries and accidents

❑❑❑

Chapter

16

GENERAL FARM BIOSECURITY

Prevention is better than cure. Bio-security is a term, which means that the collective measures taken to keep the disease away from the population. All preventive measures taken for the prevention of disease conditions will hold good more for Ostrich rearing unlike in any other species. These birds are known to have amazing immunity against common ailments. Still, the places where these birds are maintained require top bio-security requirements, especially the chicks. Farm sanitation is part of bio-security; which can be effectively carried out only when the farm is located in a proper place with scientifically designed poultry houses and equipment, providing locational, structural and operational biosecurity. The farm should always be kept clean, so that the problem of diseases will be minimized, which in turn reduces the losses due to mortality and morbidity and improve the production performance.

Diseases are spread in many ways like

- By direct bird to bird
- By visitors, farm personnel, workers, etc
- By footwear or gloves from one section to another section
- By contaminated equipment from one place to another
- By vehicles, etc

What must be done to reduce the spread of infection in the farm

- Feathers and other waste materials must be removed from the farm premises at least once in a week and burn them. The manure pit, incinerator and burial pit should be located as far away as possible (not less than 30 m) from the Ostrich houses.
- Footbath using phenyl, lime powder, etc. at the entrance of the farm and at the door steps of each house should be maintained
- The farm area should be totally fenced to prevent trespassing and entry of outside persons, stray dogs, livestock, poultry etc.

- The legs should be dipped in the foot bath and legs and hands washed up to knee and elbow respectively, before entering the farm house.
- Always youngest birds should be attended first and then other age group of birds.
- Separate set of equipment for each house and batch should be used.
- Empty feed bag should not be brought back from the poultry house to the feed room. Recycling of feed bag should be strictly prohibited.
- The farm gate and house doors should be closed to prevent the birds from coming out and to screen entry of wild birds, dogs, rodents etc. into the poultry house.
- Entertaining guest and visitors to the poultry farm should be avoided.
- Fomites in brooders, feeders, electrical fittings and other inanimate objects especially curtains will harbour various disease-causing microbes. Therefore, while cleaning and disinfecting the Ostrich house and equipment, the fomites also should be cleaned and disinfected thoroughly.
- As far as possible the use of curtains should be minimized in tropical open-side Ostrich houses; because they not only harbour disease producing organisms; but also block the ventilation.
- The farm wastes should not be disposed of in the waterers around and near to the Ostrich house; but dispose them off safely into the drainage channel, manure pit or burn them.

General cleaning and disinfection of Ostrich houses and equipment

Proper cleaning and disinfection of Ostrich houses and safe disposal of farm waste are essential for maintaining optimal health status of the birds, which in turn results in better production performance, lower mortality, and high net profit per bird. Ill health and sub-clinical infection will not only cause heavy mortality but also causes morbidity, poor production performance and severe economic losses.

- **Cleaning**

"Downtime" is the period between two batches reared in the same house. Duringthis period the shed should be empty, where no birds are reared in a poultry house and will be utilized for cleaning and disinfection of the house. Soon after depopulation, all the temporary fixtures like feeders, waterers, brooders, etc. should be removed from the poultry house and the litter is removed. While cleaning and removing the litter, dustiness should be avoided and/or the curtain is closed, so that infection will not spread to the neighbouring pens/ houses through the dust. If the litter is too dusty, water must be sprayed containing disinfectant over the litter, before lifting the litter totally. The feathers, litter around poultry houses (outside) should be removed at least 10feet away. The floor should be scrubbed and cleaned, from wall to bottom of the roof (ceiling), wire-mesh, curtains, water and electrical fittings and other immovable fittings like cages, to make them free of dust, cob webs, litter, droppings etc. Flame gun may be used to clear and clean the room from fine dust

- **Washing**

If available, jet of water or steam under high pressurecan be used for washing all the fixtures. To remove the dried stains or droppings on the wall, floor, cages and equipment, they may be soaked in water or1-2% caustic soda for some time, then scrubbed well using a hard bristle or metallic brush. Later, the house may be sprayed including the ceiling ,mesh, walls, floor, outer four sides up to 10 feet, curtains, roof etc., with disinfectants like 2 % black phenyl, 2% lye (sodium hydroxide), 2% Lysol, 4% Aldepol, 5% formalin, 4%Attack, 0.25% Disfect-S, 0.5% Hypochlorite, 1% Savlon ,50–100 ppm chlorine solution, 100 ppm Iodophor, 500 ppm Quaternary Ammonium Compounds or any other standard disinfectant, at the recommended concentrations.

- **Fumigation**

Fumigation is one of the indirect methods of disinfection. For more efficient fumigation, the Ostrich house should be pre- cleaned (dust-free), room temperature should be above 30°C, relative humidity should be more than 70% (for this spray water over the floor) and the side curtains and doors should be closed, so that the formalehyde gas will be retained in the house for a longer time. The fumigation may be preferably carried out after arranging the brooder sets but at least 12 hours before the arrival of chicks, so the house and equipment remain practically sterile, by the time the chicks arrive in the farm. However, fumigation is not safe and advisable, if other batches of chicks are present in the neighbouring pens, without any solid wall partition in between.

- **Water quality management**

Water is the most important and abundantly required nutrient for commercially reared Ratite species. Continuous supply of clean, fresh and wholesome potable water is highly essential. Most often the farmers neglect to provide good quality water to their birds and end up with complicated and multi various disease, management and nutritional problems. Contaminated water is largely responsible for a number of problems in commercially reared birds causing poor performance, increased morbidity, mortality and in numerable water borne disease.

The potable water for birds contain three major types of contaminants namely, minerals, microorganisms and organic matter, including pesticide residues. The mineral levels in water decide the total dissolved solids, salinity and hardness. They are mostly depending on the mineral levels in the soil and show only minor fluctuations in their levels, based on the season and water table. During summer, when the water level in the well goes down, the mineral concentration in the well water also goes up. On the other hand, during monsoon the mineral concentration in the well water is invariably lower, due to raised water table and dilution with water. At the same time, the microbial load and organic contaminants in the water shoot up due to rain and flood water. To avoid flood and drainage water, the farm must be located in an elevated place. The microbial load and organic matter in the water fluctuates greatly, depending on the season, source of water, storage conditions, personal hygiene and type of waterers.

Faecal contamination of water will exhibit the presence of Coliform organisms, nitrates, nitrites and ammonia. A safe drinking water for Ostrich should be free from Coliform organisms at primary source and not more than 10 Coliform organisms/100 ml at the drinking end (waterers). *E. coli*, Salmonella, Campylobacter, Arizona and Klebseilla species are the common bacteria present in the contaminated water. Presence of organic matter and bacterial contamination even alter the chemical composition of the drinking water, due to putrefaction of the organic matter by bacteria, resulting in change of pH, increase in the phosphate content and by formation of hydrogen sulphide.

Water requirement

Ostriches require 2.3 grams of water for every / gram of dry feed intake. The production of white urine by Ostriches can be indicative of illness or water deprivation. Ostriches can survive long periods in the wild without water by drinking cumulatively large quantities when it is available. Restriction of water under artificial, intensive rearing system leads to a decrease in dry feed intake. Therefore, water should be provided ad libitum during day light hours. Four troughs of water should be kept in the shades in the paddock to keep the water cool as Ostriches will avoid drinking excessively warm water.

Sanitation of drinking water

Filtration, sedimentation, ionization, ultraviolet treatment and chemical treatment are some of the methods to purify the water. But chlorination of water is still the best and cheapest methods to get rid of micro-organisms. But removal of excess dissolved minerals by cheaper and simpler methods is not practicable. Hence, the farmer should change to other water sources in case of excess minerals in the water. Birds are relatively more resistant to Fluorosis, than cattle and other live stock. Addition of chlorine to water also reduces oxidation of iron, thus reducing the formation of rust in the pipes and waterers. Five grams of stabilized bleaching powder, with about 35% available chlorine, should be added to 1,000 litres of drinking water, in order to maintain a chlorination level of 1 to 2 ppm. A minimum contact period of 60 minutes should be given, before offering the water to birds. However, over night contact time is preferable. A 50% hydrogen peroxide (H_2O_2) solution at 1 ml / 50 lit. of water is a cheap and better alternative to chlorine for water sanitation. It also needs shorter contact time.

- **Cleaning of water tanks and pipelines**

This is very essential to supply safe drinking water to the birds. The water tanks should be always covered with an air tight lid, to avoid access to wild birds and rodents. Cleaning and disinfecting the water tanks and flushing the pipelines once in a month is advisable. This will not only reduce the microbial load; but also removes the clogs in the pipeline due to algal, fungus or any other solids or dust. For this purpose, a sanitizer or a disinfectant will be effective against algal and fungi also. Same sanitizers (chloride, hydrogen peroxide) at higher dose will destroy the algae and fungi, but such water shall not be supplied to the birds. Hence, such flushing and cleaning of pipes have to be done during night time or when the sheds are empty.

- **Air**

Many of the diseases are usually air borne. Ostrich farms, as informed earlier should not be located nearer to any poultry farm or public roads. Small particles of contaminants from extraneous source may be carried into the farm and cause significant damages. When laying out a farm, these points should be considered. Since, the younger age groups of birds are more prone, care should be exercised that chicks are not in the downwind direction from older birds.

On farm bio-security will be successful only if all the personnel working in the farm are well trained in these vital aspects. Diseases are easier and economical to prevent rather than to treat. A total commitment from all staff of the farm is the only way to protect farm and ensure total bio-security.

❑❑❑

Chapter

17

TRANSPORTATION OF OSTRICHES

Ostriches are mostly transported for breeding, rearing and slaughter purpose. The birds are not advised to be transported for treatment purposes. The purpose usually indicates the type of arrangements required for transportation. Ostriches can be safely transported even up to 3000 kms, if required comfort is provided. These birds are usually transported as day old chicks, three months, three or six months old, six to 12 months, yearlings or 3 to 4 years (for breeding). Each of these different age groups has different requirements, and specialised transport systems are therefore required for each group. However, it is advisable to transport few expert workers along with a veterinarian during transportation of juvenile or adult Ostriches. The birds are always accompanied by relevant documents informing the place of origin, place of destination, health certificate, sales invoice, etc.

Ostriches have the instinctive flight response. It has been our experience, that the birds can be managed as a group rather than as individual. Young Ostriches have strong flocking instinct, which can result in flocks of young birds running into wire fences if an alarm is triggered. As they grow, this flocking instinct diminishes, and adult birds tend to react individually when alarmed. However, the flocking instinct of Ostriches can be useful when they are being loaded or unloaded, or are required to go through raceways, because the birds will follow each other once a lead bird has been enticed forward. Manual restraint can be dangerous, both for the birds and the handler. The birds should be normally trained to get used to handling by the handler. Younger birds till the age of six months can be manoeuvred with less strain. Capture myopathy is the result of improper handling and planning. In most cases these conditions proves to be fatal.

Loading into the vehicle

Use of properly designed trailer or vehicle is most important to the successful and safe transportation of Ostriches of all ages. While it is a common practice to use a converted horse or cattle vehicle, this is accepted only if the vehicle meets minimum standards. The vehicles should be fully enclosed; however, birds should never be transported in an open vehicle too. The vehicle must have no gaps in which a bird could catch a foot or neck, should have no sharp edges and have enough ventilation and good air flow. In extreme climates, birds should not be transported in hot parts of the day. A non-skid loading ramp is essential, as it makes loading and offloading easier and safer, and eliminates leg injuries. A darkened interior is also beneficial as it simulates night and encourages birds to bend down during the move. When transporting Ostriches, the headroom should be more for free movement of head to avoid head injuries. After every use, the vehicle should be cleaned thoroughly and disinfected. This procedure will greatly reduce the chance of cross contaminations from farm to farm.

There are only general guidelines for loading and unloading ratites. It is best to load during the day, however, if absolutely necessary birds can be loaded in night hours. More important than technique is the handler's attitude. The process should never be rushed. It should be slowly and gently; the birds will be loaded into the vehicle one at a time. Chicks are usually carried on board while older birds are normally walked directly in to the vehicle. Some bird will easily load by backing them into the vehicle. The standard area required for individual birds being transported for breeding is 10 x 10 ft. This place is sufficient for transporting the birds for two to three days or even more. Birds have been observed to respond to regular handler rather than the strangers. Loading of adult males especially during the breeder season are most dangerous among different age groups. While loading the birds in groups, old group should be maintained and should not be mingled with new member or new groups. The floor is usually laid with 5 to 6 inches of good river sand and preferably covered with thick cover.

Transporting the birds

As soon as the birds have been loaded into the vehicle, the vehicle should depart immediately. Once the vehicle begins moving, birds will usually sit down. When it comes to a stop, some birds immediately stand up. Therefore, any sudden starts or stops could cause birds to fall and injure themselves or other birds. Hence, the process should be gradual and slow. During the move, the vehicle should periodically stop and the birds are inspected visually to ensure that they are doing well. Depending on the distance that the birds will be transported it may be necessary to stop and offer water and feed to the birds. Only a small percentage of bird will drink or eat, primarily due to stress, but it is important that the birds are provided with drink or eat during the whole period of journey. Younger birds are more inclined to eat and drink that are older birds. Ostriches may be tranquilized, if needed for transportation. However, this is an uncommon practice and it is no substitute for proper handling and the use of an adequate vehicle. Avoid transporting breeding birds during their laying season. The sudden stress brought on by the move may adversely affect the birds and they may cease the laying activity completely. The birds should be provided with good ventilation throughout the journey period.

Unloading

Prior to unloading, it should be ensured that the birds are given ample time to get on their feet and establish their bearing before they are slowly and gently escorted off the vehicle. Because of the physical stress during transportation birds generally become very easy to handle during unloading.

It is necessary to unload the birds at night; extra precautions are to be taken to ensure that the birds do not run into anything when exiting the vehicle. If possible, the area needs to be illuminated, in which the birds will be held. Once the birds are unloaded, they should be assisted in locating feed and water. Unloading in darkness has no lasting harmful effect on birds. The effect of keeping them confined unnecessarily in a vehicle is much more detrimental to them and dramatically increases the likelihood of a birds succumbing to stress or injuring another bird.

Air transport

Ostriches may be transported via air depending on their age and or size. They can be transported as young as one day old in some cases. Chicks from one to thirty days old are normally placed in air carriers while older birds are sometimes placed in individual compartments. Air transportation is viable alternative to ground transportation when eggs or birds have to be moved rapidly within a specific time frame.

As this industry is growing, more information on transportation of Ostriches will come to the limelight. The major issue to be considered is safe movement of Ostriches from one place to another keeping in mind the welfare of the birds.

Transportation of ostriches

(Courtesy:PGRIAS, Kattupakkam, TANUVAS, Tamil Nadu, India)

Chapter

18

OSTRICH SLAUGHTER AND THEIR PRODUCTS

Ostriches are slaughtered usually at 14 months of age in South Africa. In some other countries like Australia, U.S, Israel, etc. these birds are considered as meat animals; hence they are slaughtered at much younger age of 9 months due to poor conversion of feed. Pre-slaughter handling of the birds is very important; in the sense it determines the quality of meat and also satisfies the bird welfare issues. Under commercial ventures, Ostrich abattoir should be built with standard plans and to be operated with good hygienic conditions. Due to the physical appearance of Ostriches, regular slaughter house for livestock or poultry is not suitable for Ostriches. They need to be specially designed for Ostriches. As observed in regular slaughter house, separate areas for holding the birds, stunning, killing, plucking, skinning, evisceration, deboning, chilling, packing, freezing and dispatch should be made available.

Once the birds have been gathered in holding pens, they are guided into the slaughter box, one bird at a time. Preferably, birds should be guided by their handler into the box as this will help them to stay calm. The slaughter box should only be large enough to contain one Ostrich. Strips connected to a chain on a mechanical pulley arc, placed around each leg and the bird is stunned with an electric stunner (similar to that used for sheep). Before the bird falls to the ground, it is pulled up by the chains around the legs and hung upside down. It is then moved to the next person in line where, the head is severed and the blood is allowed to drain. It is plucked and skinned after which the carcass is removed and normally hung by the wings and eviscerated.

The eviscerated carcass is allowed to cool in the chilling area at 1 °C. The chilling period varies in length between countries and operators from only a few hours to a full 24 hours. Deboning is carried out and the meat is sorted into different cuts, normally vacuum packed in 2-kg packs and either dispatched fresh (with a shelf life of approximately three weeks) or moved to the freezing area for storage at a temperature of -20 °C.There is no difference between meat quality of male and female when, birds have similar age and weight. Parts yield may reach up to 42%, depending on genetic background, rearing system, and slaughter techniques. As the Ostrich is a runner bird, the absence of keel in the sternum and the reduced muscle mass in the wings, providing it with a reduced breast muscle mass, and therefore saleable carcass parts are concentrated in the posterior limbs. Carcass utilization is mainly limited to the back, thighs, and legs. Parts correspond to each muscle that composes these regions.

Flow Chart of Ostrich slaughter

Holding of birds
↓
Stunning
↓
Bleeding
↓
Plucking
↓
Removal of head
↓
Skinning
↓
Hanging of bird in wings
↓
Removal of legs
↓
Opening of sternum
↓
Evisceration
↓
Back fat removal
↓
Removal of legs and thighs
↓

Thighs and legs	Other parts
↓	↓
Refrigeration	Removal of neck
↓	↓
Deboning	Removal of breast fat
↓	↓
Packing of meat	Processing of meat

Ostrich products

Commercial Ostrich results in four major products from these birds. Earlier only three products were taken into consideration via meat, skin and feathers. Now, with improved information available on the use of Ostrich oil, this is also fetching good revenue from the pharmaceutical industry. Meat and skin are fetching standard revenue for the industry. Revenue from feather industry has reduced.

Feather

Feather fetches only about 5 to 10% of total revenue from the birds. But this paltry amount can make a difference between profit and loss in tight situation. Though the market for the feathers has reduced, but still nothing can replace the uniqueness of this feather. The wing and tail feathers of young Ostriches can be harvested. Birds of above 60 kgs body weight are ideal for feather harvesting. Immature or green feathers are never to be harvested inside the farm premises. Only mature and ripened feathers need to be harvested. The feathers are normally harvested during the rest phase of the breeders. The feed given to the bird should be well balanced in terms of quality and quantity. Feather harvested should be separated as male and female feathers. And within the same bird, feathers from different parts of the body must be stored separately. The storing time of feather should be minimised as far as possible. The different types of feathers harvested are wing quills, Byocks, Long upper wire coverts, Lower wing coverts, tail feathers and sides (bodies)

Income from feather depends largely on the physical condition and quality of feather. Damaged or soiled feathers will never fetch any price. In terms of quality, some traits such as width, symmetry of feather, size and appearance, length of quills, are to be well taken care of in the marketing. Feather of South African Black are better Zimbabwe Blue Ostrich due to different genetic makeup.

Ostrich egg shell decorated

(Courtesy: PGRIAS, Kattupakkam, TANUVAS, Tamilnadu, India)

Leather

Ostrich leather has its own market. And the market is of premium quality. The uniqueness of the leather is its follicles. Feather follicles are very unique to these birds. The skin may be removed either from the dead birds or during slaughter. The following care must be exercised while handling the skin of these birds;

- Skin should be removed from the birds as soon as possible
- The carcass should never be dragged by its legs, as it leaves irreversible marks of the skin
- The feathers should be totally plucked before the removal of skin
- The skin should be trimmed of the meat and fat and washed well for blood stains
- The skin should also be delivered as soon as possible. Otherwise, it is salted thoroughly on the meat side and preserved for short period. The applied salt should be of equal weight.
- Skin should be stored in a cool place to prevent decomposition

Quality of skin

Skin is one of the major revenue provider in the whole business of Ostrich rearing. Utmost care should be given to minimise the damage to the skin either knowingly or unknowingly. Skin size, visible skin damage and follicle development determine the price of the skin. If the expectations of the buyer are not met, a huge set back in terms of financial loss can be expected. Below diagram indicate the actual Ostrich skin as it is expected in terms of appearance.

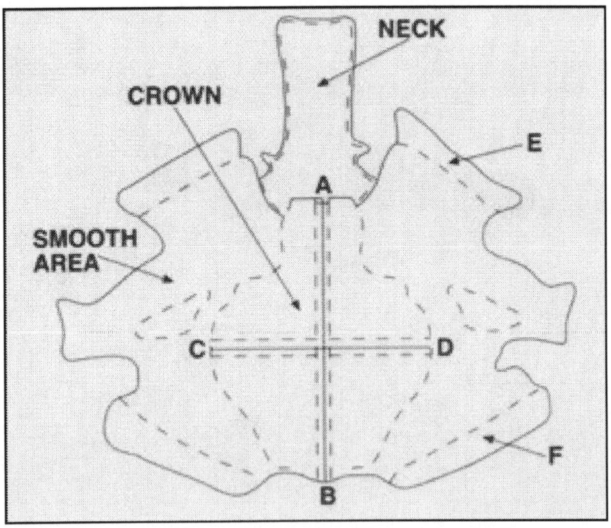

Skin size

Skins are given different grades based on their size. The size of the skin varies from minimum 80 square decimetres to more than 150 square decimetres. Areas of skins for different size groups are as follows:

- A+ = 156 + square decimetres
- A = 130 to 155 square decimetres
- B = 120 to 129 square decimetres
- C = 100 to 119 square decimetres
- D = 80 to 99 square decimetres

It is presumed that a bird reaching a weight of at least 95 kgs and above will have a minimum 140 and above square decimetres and fetch high price.

Damage

The price will be determined by the physical visible damages to the skin, especially the crown area of the skin. The crown as mentioned earlier is the centre part of the skin where the maximum feather follicles occur. The position of damage in the crown area determines the grading of the skin, because it affects the cutting value of the skin. The damage increases with the age of the birds.

Follicle development

Skins are also evaluated based on the follicle developments. Though the exact way of assessment of the follicles is debatable, but still the important aspect seems to be a well-developed rounded follicle size. The size increases with the age and the follicles needs to be matured enough to be thick and steady.

Ostrich leather products

(Courtesy:PGRIAS, Kattupakkam, TANUVAS, Tamilnadu, India)

Meat

Meat yield varies according to the age at slaughter. A 12-month-old Ostrich weighing 100 kg live weight will yield a carcass weight of 60 kg (a dressing percentage of 50 –

60 %). Depending on age, sex and management, the dressing percentage ranges from 56 to 64 percent. Other non-meat component yields 36 to 44 %. Males yield approximately 1.5 to 2.0 percent more than females. The legs yield about 75% of carcass weight, which forms the bulk of meat. An average estimation reveals 60% of lean meat, 25 % of bone and 15% fatin older birds, the yield of fat and bone increases. For marketing purpose, this meat is classified as prime cuts and trimmings in the ratio of 1.5:1.

Meat Products;

Traditional delicacies of Ostrich meat are dried sausages and biltong. Commercially, these meats are marketed as Jerky, fillets, bacon, ham, pastrami, meat cubes, etc.

Meat quality

Ostrich meat is lower in fat, compared to other animal meat. As a result, one can cook it quickly. Ostrich meat also has low fatty acid. As per USDA, the comparison of the Ostrich, chicken, emu, turkey, pig and cattleis given below as illustration;

Comparison of Ostrich meat with other meats

Species	Protein (%)	Fat (%)	Calories (Kcal)	Iron (mg)	Copper (mg)	Mg (mg)	Vit B12 (mcg)	Cholesterol (mg)
Ostrich	26.9	3.0	175	3.2	0.14	22.6	5.7	83
Emu	28.4	4.7	164	5.0	0.24	28.7	8.5	87
Chicken	28.9	7.4	190	1.2	0.09	15.8	0.40	89
Turkey	27.4	5.0	235	1.8	0.09	24.0	0.33	102
Cattle	25.4	16.4	255	2.4	0.09	21.0	2.0	86
Pig	29.3	9.7	212	1.1	0.15	19.0	3.9	86

Fat/ oil

The fat is the thickest on the back of the bird and is removed from pulling it away with one hand while cutting the carcass with other. It is immediately washed and kept in clean vessel. The yield varies with the age, species and nutritional status of the birds.As far as possible, fats harvested from the carcass should be refrigerated immediately. Ratite oil is semi solid white mass at 200°C, while heating at 6000°C, they are practically yellow colored liquid. They are insoluble in water and their water content is less than 1%. Other properties are in the range of

Specific gravity – 0.94 g/ml
Refractive index – 1.45
Saponification value – 194.1
Iodine value – 70.97
Acid value – 0.64
Peroxide value – 0.60

The composition of Ostrich oil is very much similar to human skin. Ostrich oil has proven effect as anti-inflammatory and is highly penetrating in nature (non-comedogenic). They are mainly effective as a topical application. Currently, these oils are more useful to cosmetic industry for the preparation of various face and body creams. The basic process involves the following steps

1. Recovery of fat during slaughter
2. Removal of dust and excess meat particles
3. Melting of fat
4. Processing of crude oil with filtration or centrifugation to remove meat residues
5. Conventional method of refining, bleaching and de-odorising
6. Conventional caustic treatment to remove free fatty acids and proteinaceous residues
7. De-odorising can also be done high vacuum steam purging
8. After the removal of metal, soaps, protein and fee fatty acids, they are stored for further use

Composition of Ostrich oil in comparison with other common oils

Type of oils	Saturated (%)	Unsaturated (%)	Linoleic (%)	Linolenic (%)
Sunflower	11	20	69	-
Ground nut	18	48	34	-
Soybean	15	24	52	6
Lard	41	47	11	1
Olive	17	77	7	2
Ostrich	**32**	**58**	**19**	**11**

Ostriches being kept on display for an exhibition

(Courtesy:PGRIAS, Kattupakkam, TANUVAS, Tamilnadu, India)

Chapter

19

FUTURE OF OSTRICH FARMING

The future for Ostrich farming is indeed bright. South Africa, who were the leader in Ostrich farming for so many years are being challenged by many European and Asian countries. Iran and China are racing to hold the second spot in Ostrich rearing. Due to severe H1N1 outbreak in South Africa, the birds are being slaughtered as a bio-security protocol. The export has been severely affected; local consumption has also been badly affected. The whole industry seems to be falling apart. South Africa is the home to 60 to 65% of total Ostrich birds in the world and has a share of more than 90% of World Ostrich products. Ostrich has been surviving the punitive conditions for decades now. It is a very hardy bird and has a huge potential to be a perfect farming venture when compared to other farming enterprises. The variety of products that can be garnered from this bird is not found in any other agro-business venture. Ostrich farming is much more profitable than raising other traditional farm animals. Ostriches are loving birds; they are quite easy to raise and require little maintenance from farmers. One of the reasons why Ostrich farming is a lucrative business is the number of valuables an Ostrich has to offer; and very little ever goes to waste.

The scenario is encouraging for the existing farmers to go for expansion and for new entrepreneurs to jump in the business. Some of the late entrants to the Ostrich business have shown promising results in Ostrich farming. Some of the Asian countries like China, Pakistan and Nepal have pumped in huge investment for future prospects. It is no doubt coined as "farming of the future", "Sun shine industry", "Bird of the millennium", etc.

The focus of Ostrich farming should now be more on the oil, skin, meat and meat products. More emphasis should be given about the quality of meat and oil. It is well known that Ostrich meat is red meat and has the lowest fat. It is almost less than 2 %. This meat is low in calories and cholesterol. The meat is as juicy as any other meat and is also rich in vitamins. Similarly, the focus on oil and their qualities should also be not lost. The oil is rich in unsaturated fatty acids and the contents of Linoleic and Linolenic acids are also quite high. It is effective against almost all ailments which a human normally suffers like, joint pain, joint inflammation, hot spots, sun burns, etc. When taken internally, it is rejuvenator and muscle relaxant. Recent studies have shown promising results about the use of Vitamin. K_2 (MK4 and MK7). It is well known that Vitamin K_1 plays very important role in the blood coagulation. The utility of Vitamin K_2 is very recent. They play a pivotal role in the stabilising the cardiovascular and bone health. Vitamin K2-7 (menaquinone) supports healthy bone mineralization by acting as a cofactor in the carboxylation of osteocalcin, thus allowing calcium to bind to this protein in the extracellular matrix of bone. It also supports healthy cardiac and vascular function. It concentrates in the tissues of the arteries, where it aids in carboxylation of Matrix Gla Protein to help maintain normal calcium metabolism and support healthy vascular elasticity, which is very essential for healthy cardiovascular activities. Additionally, it also supports healthy lipid levels and glucose metabolism. These benefits are of recent origin. More information in this subject is available in the following websites;

- **https://nutrition.basf.com/global/en/human-nutrition/products/vitamin-k2.html**
- **https://www.healthline.com/nutrition/vitamin-k2**
- **https://en.wikipedia.org/wiki/Vitamin_K2**

The findings that ratite oil (both Ostrich and emu oil) are very rich in Vitamin K_2 has renewed the interest of the Ostrich growers. It is envisaged that the market size of Vitamin K in the following days will have a CAGR of not less than 32% **(https://www.medgadget.com/2020/04/vitamin-k2-market-size-is-projected-to-reach-290-3-million-by-2026-industry-players-dupont-health-and-nutrition-dsm-kappa-bioscience-basf-kikkoman-corporation-and-others.html).**

This it is enough for the Ostrich growers to reactivate their activity. Ostrich provides variety of products. The "wonder oil" has been found as Vitamin K which makes bone and heart healthier. More information can be read in **https://i.pinimg.com/originals/5d/58/b1/5d58b164dd46b54f677b6680e7f27d06.png.**

How to make Ostrich farming successful

As in any farming enterprises, Ostrich farming also has many issues and concerns. These concerns can be addressed successfully. This industry has seen many ups and downs and has always been resilient and robust. It has shown more peaks every time that it has made a comeback. The industry has survived for more than a century under the modern times. The major constraints are summarised and the newcomer in Ostrich enterprises need to take care of these issues before investing in the farm;

1. The farming needs to be endorsed by the government. No farming / enterprises would ever succeed, if it does not have the backing or support of the local government.
2. All means to keep the production cost as low as possible. Feed accounts for more than 60% of production cost. Hence, the feed requirements for Ostriches need to be kept at the lowest possible cost. Availability of cheap labour is an added advantage.

3. Co-operative activity is more preferable than as individual activity. Since, this farm unlike other farming enterprises needs strong cooperation / integration. The different sectors of the Ostrich farming need more concentrated service to succeed. The cooperative society undertakes all activities from advertisement to start Ostrich farm to marketing of Ostrich and Ostrich products.
4. All the basic information regarding the Ostrich production including hatching, chick management, grower and breeder management, feed and feedings, health cover, etc must be available in public domain. The basic needs like feed, eggs, incubator, hatching needs, technologies, tanneries, etc should be available in the market.
5. Ostrich meat is a red meat; hence the price needs to be similar to the cost of other red meats. Ostrich meat pricing cannot be same as poultry meat.
6. Public awareness on the goodness of Ostrich meat should be taken care of
7. Agra-touristic activities are also a good option to market the Ostrich products.
8. The Ostrich Association/ Society formed locally should take care of the interest of its members first. They should conduct regular National/ International trainings, workshops, conferences, public seminars, fairs, exhibitions, etc to keep themselves and the public update on Ostrich rearing. They should also have close collaborations with Government institutions, research institutions, market, etc.
9. Regular promotion of Ostrich and Ostrich products will keep the interest alive in public domain. All the operations should be integrated well and coordinated to achieve the desired results.
10. As per **http://www.journals.zu.edu.eg/journalDisplay.aspx?Journalld=1& query Type= Master**, Ostrich farming is successful because of Cheap feed and Nutrition, Hatching and health coverage.

Is Ostrich farming viable?

The current trends suggest that Ostrich farming is surely a viable option and the stock position of Ostrich in earlier unknown countries is increasing. The farmers have come to comprehend the recompenses of Ostrich farming in the long run. It is one of the best available options that provides sustained income and also provides variety of products. However, the following needs special mention and attention for the existing growers;

1. Optimal slaughter age
2. Feed and feeding cost
3. Age at sexual maturity
4. Health coverage
5. Fertility and hatchability
6. Mortality especially the chick mortality from 0 to 12 weeks
7. Genetics of the bird

8. Marketing of meat and their products.
9. Purpose of rearing (whether meat or breeder farm) and
10. An eye on worldwide Ostrich research verdicts.

One of the major reasons for the failure of Ostrich farm worldwide is the overestimation of market demand and the farmers / growers entering into the venture with inadequate knowledge base, infrastructure and institutional support.

Visit to an Ostrich farm in Australia

Meeting the other Ostrich researchers at an international conference.

❑❑❑

Chapter

20

FREQUENTLY ASKED QUESTIONS (FAQ)

1. *What are acceptable climates for raising the Ostriches? Will they survive in any climatic conditions?*

   Ostriches are very adaptable to many climates and temperatures. Ostriches are basically from South Africa and the climatic conditions are very much similar to our country. They can withstand temperature from - 5°C to + 50°C.

2. *Are Ostriches aggressive?*

   They can be at times very aggressive, but Ostriches are not naturally aggressive until they feel trapped or cornered. It is highly unusual that an Ostrich will attack a person but there have been instances where the males will get somewhat aggressive when trying to protect the eggs. Most aggressiveness is seen when the Ostriches fight amongst themselves, and during breeding season.

3. *At what age does an Ostrich hen begin laying eggs?*

   It is difficult to answer this question because they normally start laying eggs when they reach 30 to 36 months. There have been instances when the Ostrich has come into laying during 4-year period.

4. *How many eggs do Ostriches lay in a year?*

   Ostriches normally lay eggs round the year but the numbers are reduced during July – November in a year. They are known to lay about 40 to 60 eggs in a year. Genetics and Nutrition play a very large part in the number of eggs laid. For optimum production, it is always best to select Ostriches with good pedigree and subsequently good feeding is also required.

5. *Will an Ostrich hen lay without a male?*

   Yes, the hen needs the consistency of mating to promote ovulation and hence a greater number of eggs are noticed during the breeding season than the off season

6. ***Are these birds seasonal in mating?***

   Yes, they are seasonal in breeding and the breeding varies from Northern hemisphere (June to December) to Southern hemisphere (December to May). Even during off season breeding has been observed.

7. ***Is there a visible way to tell male and female in a flock?***

   Yes, the males would have black plumage covering the body with white terminal feathers and the female birds would have grey colored body feathers with terminal white feathers.

8. ***Can we consume Ostrich egg and meat?***

   Yes, the eggs and meat can be consumed just like chicken egg and meat. The egg consistency is slightly rubbery than chicken eggs and the meat taste like a beef meat. Fat percentage in meat is lowest among all the red meat (1 – 2%)

9. ***What does the Ostrich feed on?***

   Ostriches can be normally fed with regular compounded feed, but the requirement of fiber and micronutrients contents in the feed are higher than chicken.

10. ***Do Ostriches graze?***

    Yes, they graze and they are very selective in grazing. Preferably high protein fodder like Desmanthus, Stylo and Lucerne can be raised for Ostrich feeding

11. ***Do Ostriches have to be vaccinated?***

    Yes, they need to be vaccinated against New Castle disease like the one followed in broiler farming.

12. ***How to identify the onset of breeding in a flock?***

    Onset of breeding can be identified by observing the males, which would reveal the presence of red coloration in the shank, beak, vent and around the eyes.

13. ***What are the salable products in Ostriches?***

    Almost all parts of the Ostriches can be used for marketing. Major products include Skin and meat and the minor products include oil, feather, eggs and egg shells.

14. ***How to restrain the Ostriches?***

    At least four well trained persons are required to restrain one Ostrich. They have to be restrained either from the sides or from behind. Ostriches mostly kick in the front side only

15. ***Do Ostriches bury their head under the sand?***

    Ostriches never bury their head under the sand, it is only a myth. They sit on the ground and stretch their neck parallel to the ground and from far it looks as if the Ostrich has buried its head.

❑❑❑

REFERENCES

1. Engelbrecht A., 2014. Ostrich Manual.
   https://www.elsenburg.com/sites/default/files/ebooks/2015-04 29/Ostrich%20Manual_English%20ed_%202014_content.pdf
2. Hastings M.Y., n.d. A HISTORY OF OSTRICH FARMING – ITS POTENTIAL IN AUSTRALIAN AGRICULTURE. Retrieved from https://www.doc-developpement-durable.org/file/Elevages/autruches&Emeus/HISTORY%20OF%20OSTRICH%20FARMING%20%20ITS%20POTENTIAL%20IN%20AUSTRALIAN%20 AGRICULTURE.pdf
3. Shanawany M. M., Recent developments in Ostrich farming. Retrieved from http://www.fao.org/3/v6200t/v6200T02.htm
4. https://www.smallstarter.com/get-inspired/Ostrich-farming-in-africa/
5. http://www.journals.zu.edu.eg/journalDisplay aspx?
   Journalld=1&queryType=Master
6. https://www.researchgate.net/publication/233647137 Ostrich farming in Poland Its history and current situation after accession to the European Union?enrichId=rgreq-da498774b59000558e3b81f60e8032ce-XXX&enrichSource=Y292ZXJQYWdlOzIzMzY0NzEzNztBUzo2NjM0MjgyMDA5MjcyMzJAMTUzNTE4NTAxNzY1OQ%3D%3D&el=1_x_3&_esc=publicationCoverPdf
7. http://www.hrpub.org/download/20180130/AZB2-11410472.pdf
8. https://www.researchgate.net/publication/323345219 Future of Ostrich Farming in Pakistan
9. http://www.fao.org/3/v6200t/v6200T02.htm
10. http://www.ltz.de/en/news/lohmann-information/20191/3 Ostrich
    Production in Germany ecological way to economic success.php

❑❑❑

INDEX

D

E

F

T

U

V

W

X

Y

Z

COLOR PLATES

(Pg. 1)

(Pg. 2)

(Pg. 3)

(Pg. 10)

EMU

OSTRICH

EMU

CASSOWARY

(Pg. 23)

KIWI (Pg. 23)

Semi (Pg. 26)

(Pg. 28)

(Pg. 34)

(Pg. 38)

(Pg. 39)

(Pg. 44)

(Pg. 45)

(Pg. 46)

(Pg. 47)

(Pg. 48)

(Pg. 49)

(Pg. 50)

(Pg. 52)

(Pg. 54)

(Pg. 55)

(Pg. 55)

(Pg. 56)

(Pg. 58)

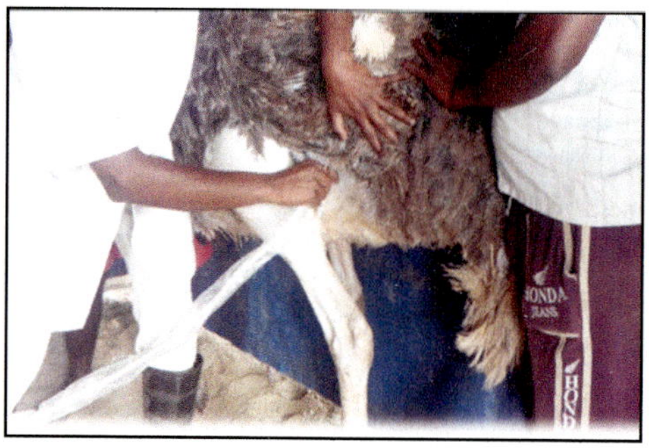

(Pg. 60)

(Pg. 63)

(Pg. 66)

(Pg. 68)

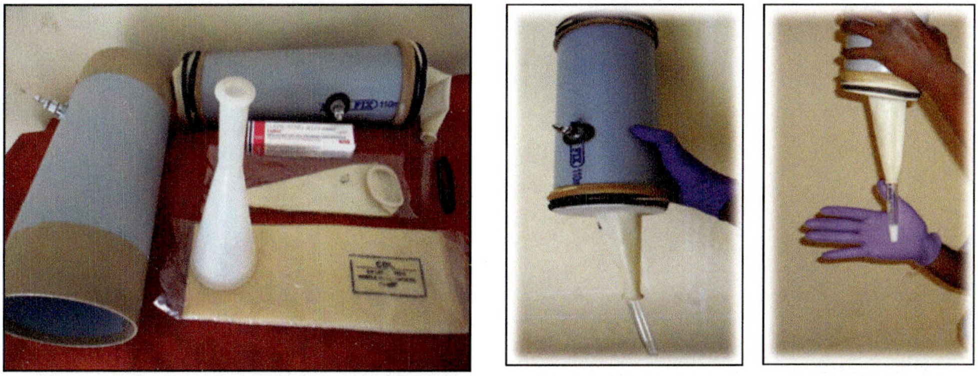

(Pg. 73)

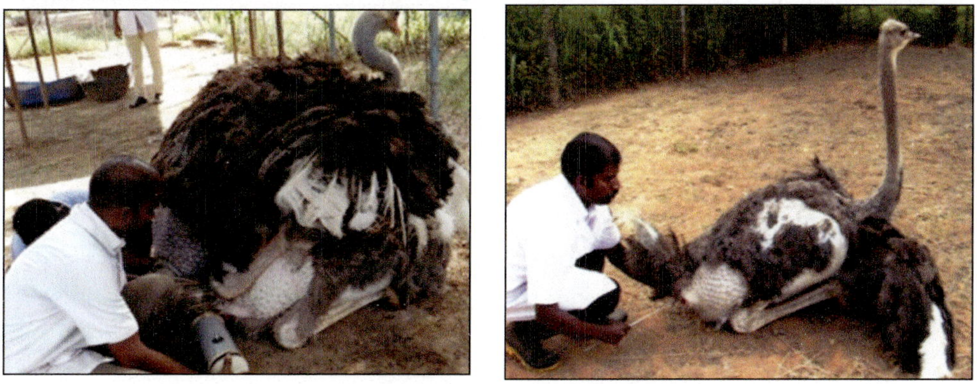

(Pg. 74)

(Pg. 76)

(Pg. 79)

(Pg. 84)

(Pg. 85)

(Pg. 89)

(Pg. 90)

(Pg. 93)

(Pg. 95)

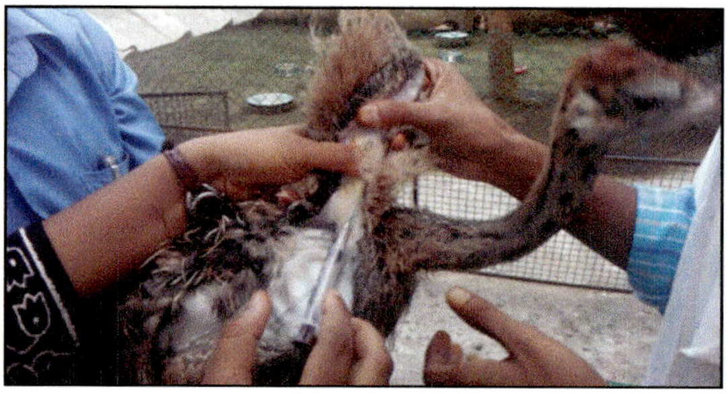

(Pg. 101)

(Pg. 103)

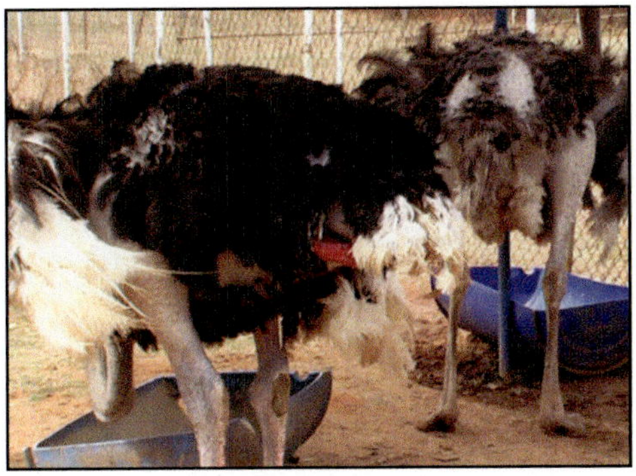

(Pg. 106)

(Pg. 116)

(Pg. 123)

(Pg. 119)

(Pg. 121)

Visit to an Ostrich farm in Australia (Pg. 127)

Meeting the other Ostrich researchers at an international conference. (Pg. 127)